W9-CUJ-320

LAND AND FAMILY IN PISTICCI

LONDON SCHOOL OF ECONOMICS
MONOGRAPHS ON SOCIAL ANTHROPOLOGY

Managing Editor: Peter Loizos

The Monographs on Social Anthropology were established in 1940 and aim to publish results of modern anthropological research of primary interest to specialists.

The continuation of the series was made possible by a grant in aid from the Wenner-Gren Foundation for Anthropological Research, and more recently by a further grant from the Governors of the London School of Economics and Political Science. Income from sales is returned to a revolving fund to assist further publications.

The Monographs are under the direction of an Editorial Board associated with the Department of Anthropology of the London School of Economics and Political Science.

Pisticci from the south

LONDON SCHOOL OF ECONOMICS
MONOGRAPHS ON SOCIAL ANTHROPOLOGY
No. 48

LAND AND FAMILY
IN PISTICCI

BY

J. DAVIS

UNIVERSITY OF LONDON
THE ATHLONE PRESS
NEW YORK: HUMANITIES PRESS INC
1973

Published by
THE ATHLONE PRESS
UNIVERSITY OF LONDON
at 4 *Gower Street London* WC1
Distributed by Tiptree Book Services Ltd
Tiptree, Essex

USA and Canada
Humanities Press Inc

UK SBN 0 485 19548 8
USA SBN 391 00288 0

Printed in Great Britain by
T. & A. CONSTABLE LTD
EDINBURGH

ACKNOWLEDGEMENTS

This book is based on a thesis written for the degree of doctor of philosophy in the University of London.

The research for it was done over twenty-one months between March 1963 and January 1966. It was financed by a London School of Economics postgraduate studentship, and by grants from: the British Academy; the Central Research Fund Committee of the University of London; the Mediterranean Research Committee of the London School of Economics, the School of Oriental and African Studies and the University of Kent at Canterbury; and from Ann Arbor, Michigan. I wish to thank all these bodies publicly.

In England A. P. Stirling inducted me to anthropology and to Italy; Lucy Mair supervised the thesis and gave me invaluable help in turning it into a book. I am deeply grateful to them.

In Italy, I was welcomed, fed, nursed, comforted, tolerated and made to understand by Rocco Mazzarone, Gilberto Marselli and Manlio Rossi-Doria: I am deeply grateful to them.

Mr John Goy prepared the index.

I dedicate this book to the protagonists, the Pisticcesi.

A NOTE ON ORTHOGRAPHY AND ALLIED MATTERS

Most Pisticcesi speak Italian as well as their own dialect, and this book contains words and phrases in both. The dialect is increasingly Italianised and is losing many of its idiosyncratic words, structures and sounds. It is written here so that, were any Italian to read it aloud, it would sound more or less like Pisticcese. To write it so is common practice.

While I doubt that any reader needs guidance in Italian, two points about the dialect may be helpful. The first is that while Pisticcesi often leave off the last syllable of words, they accentuate the remaining syllables as they would were the word complete. So, Italian scappa·re becomes Pisticcese scappa·, written scappà: it is the infinitive form and is quite distinct from sca·ppa, which is Italian third person singular, present indicative. This procedure is common to many southern dialects.

The second point is that the feminine ending in the singular and plural is an open, unstressed *e*, as in unstressed *the* or as final *e* in French verse and song. Some writers transcribe this simply as *e*; others, whom I have largely followed, use an apostrophe (e.g. *Zi'* aunt, aunts) to distinguish the dialect from the Italian feminine plural.

Another matter: Statistical Table can be translated into Italian as either *tavola* or *tabella*: the corresponding abbreviations are *tav.* and *tab.* I have not standardised my usage, and follow that of the various sources I cite.

Finally, the names of most of the people and all the places are real names, not invented ones. A few incidents are scrambled and thereby, I hope, disguised.

CONTENTS

TABLES

FIGURES

MAPS

PLATES

(Plates 1–4 between pages 62–3)

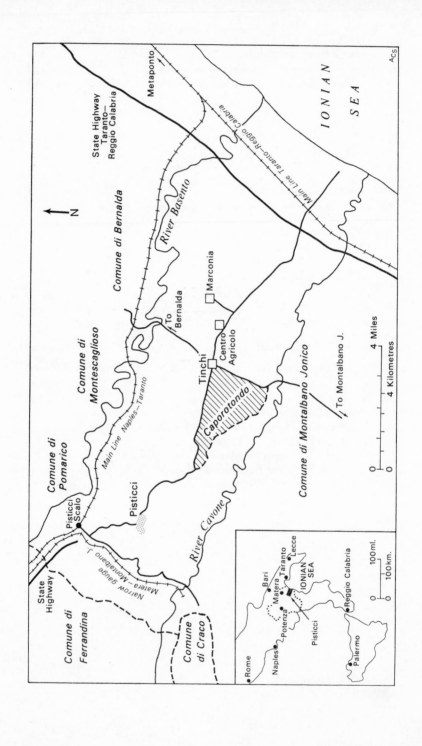

N

Comune di Bernalda

Comune di Montescaglioso

Comune di Pomarico

State Highway Taranto–Reggio Calabria

Metaponto

River Basento

Main Line Taranto–Reggio Calabria

IONIAN SEA

To Bernalda

Marconia

Centro Agricolo

Tinchi

Caporotondo

Comune di Montalbano Jonico

To Montalbano J.

Main Line Naples–Taranto

Pisticci

Pisticci Scalo

Narrow gauge Matera–Montalbano J.

State Highway

Comune di Ferrandina

Comune di Craco

River Cavone

4 Miles

4 Kilometres

0

0

Rome

Naples

Potenza

Bari

Matera

Taranto

Lecce

IONIAN SEA

Pisticci

Reggio Calabria

Palermo

100 ml.

100 km.

0

0

ACS

I

Introduction

The closer a society is to our own the more difficult it is to begin an anthropological account of it. A description of a Polynesian island can begin quite easily:

Almost before the chain was down, the natives began to scramble aboard, coming over the side by any means that offered, shouting fiercely to each other and to us in a tongue not a word of which was understood by the Mota-speaking folk of the mission vessel.[1]

Even less romantic descriptions of the ethnographer's arrival[2] tell us quite clearly that the society is exotic: the scene is set, and the real business can begin. Although south Italian peasant societies can be described in ways which make them seem exotic, they are not essentially so. Pisticci and the towns like it are mixtures of the bizarre and the 'ordinary'. The Italian normal, the local idiosyncratic, the European commonplace are there combined. Like other towns in south Italy, Pisticci has many modern aspects: it is part of an industrial nation and of a parliamentary state. Because Englishmen have travelled in Italy, met south Italians, read about them in popular weekly papers, their impressions of them are likely to be much more concrete than they are of the Nuer or Tikopia. This introductory chapter is an attempt to set bearings, to give the cultural and structural co-ordinates of the society, particular aspects of which are discussed in greater detail in the rest of the book.[3]

I

The population of Pisticci has more than doubled in the past century. For every hundred people there in 1861, there were two hundred and twenty-five in 1961.

---

[1] Firth (1957) pp. 1-2.
[2] Evans-Pritchard (1940) has an epigraph from Isaiah.
[3] Some data on climate and natural environment are given in Appendix I.

The nearly 15,000 people in 1961 live in four sorts of settlement. The town itself had a total population of 12,892 in 1951. There were three hamlets with a total population of 363, and another

TABLE 1. Population of Pisticci, Basilicata, south Italy, 1861–1961

Year	Pisticci (Absolute)	(Index)	Basilicata (Index)	S. Italy (Index)
1861	6,597	100	100	100
1871	7,737	117	104	—
1881	7,989	121	106	112
1901	8,272	125	100	125
1911	10,319	156	96	133
1921	10,388	157	93	141
1931	12,083	181	103	143
1936	11,560	175	109	149
1951	14,796	225	124	173
1961	14,847	225	—	182

Sources:
(i) For Pisticci – ISTAT: *Communi e la loro popolazioni*, 1861–1951, and ISTAT: Census 1961.
(ii) The index for Basilicata is taken from: *Atti della Commissione Parlamentare di inchiesta sulla disoccupazione*, Rome, 1953, vol. 3, p. 9, Tav. 1.
(iii) The index for S. Italy is based on figures in G. Galasso, 1965, p. 423, Tab. 5.

Notes:
(i) There was no census in 1891, to save money; nor in the second world war. The 1936 census was the first of a quinquennial series which was later abandoned.
(ii) The figures are given for residents – all those whose legal residence is in the area censussed. During this period those actually present at the time of the census vary between 95 per cent and 98 per cent of the residents.

222 people lived in 'rural centres', usually farms with labourers' cottages. Another 1,319 people lived in isolated houses – the *case sparse* of the census. By 1961 this pattern had changed, largely as a result of the Agrarian Reform Board's policy of building houses in the country and requiring people to live there.[4] (See below, Chapter 2.) The figures for 1951 and 1961 are set out in Table 1, and more detailed figures for 1963 are in Appendix II.

[4] The requirement, however, led people to make false declarations: some settlers in 1965 lived permanently in town; others left their families in town. A greater number, while living more or less regularly in the country, nevertheless regard town as 'home'. Probably about one-quarter (75 families, 350–400 people) come into the first two categories. The official census thus overstates the number of 'dispersed' people by about 20 per cent.

TABLE 2. Pisticci: Settlement, 1951 and 1961 (residents)

	1951	%	1961	%
Total population	14,796	100·0	14,803	100·0
Town:				
Pisticci	12,892	88·0	11,469	77·0
Hamlets:				
Marconia	203 ⎤		246 ⎤	
Centro Agricolo	83 ⎬	2·5	105 ⎬	3·5
Tinchi	77 ⎦		162 ⎦	
Scalo	—		47	
Rural centres	222	1·5	132	1·0
Dispersed	1,319	8·9	2,669	18·0

Source: ISTAT: Census 1951, 1961.

Marconia and Centro Agricolo, about a mile apart and twelve miles from Pisticci, are the remnants of an isolation camp built before the war for politically suspect persons. Both have sewers, and Marconia is to have running water. Each has a tobacconist, Centro has one bar and Marconia has two, as well as a butcher, a baker and a post office. Centro remains more or less as it was when it was built – a collection of huts round a courtyard, near the road. There are palm trees and a flagstaff, a small chapel which is opened occasionally. The huts are rented out by the commune at 400 lire per annum for each square metre. There were twenty families there in 1965, one of them a political prisoner's Neapolitan widow and her children. The rest are families which have access to land in the district, and no town house.

Marconia has been almost completely rebuilt. Blocks of apartments have been put up to house evacuees from the quarter of the town (Rione) which is liable to landslides. A central square, wide and open, is surrounded with colonnades and shops all built in red brick. The church was consecrated in 1959, but there is no cemetery. Building continues as more houses are declared unfit for habitation in the town up in the hills. Marconia is designated 'the new Pisticci', and speculation in building land has already begun there. To serve the inhabitants of the village and the Reform Board's smallholders in the surrounding district there is a sub-office of the communal administration, which is visited daily by a functionary and less often by an elected member of the council, a member of the executive committee. Marconia has two schools;

one is a section of the Pisticci lower secondary school, which is visited by teachers from Pisticci and directed from there, the other is a technical agricultural school which issues a school-leaving certificate in various branches of agriculture.

Tinchi is a small service village which has grown up in the last fifty years at the crossing of the road from Montalbano to Bernalda with the road from the coast to Pisticci. It serves the rural area which surrounds it, and has three grocers, a bar, two carpenters, an agricultural machinery mechanic and a barber. There is an elementary school with three teachers, a house for the *cantonieri provinciali* (the men who maintain provincial roads), and their families. There are also a combined mill, oil-press and tobacconist, and a haberdasher.

Pisticci Scalo, the town's station, is the point where contact between the traditional and the modern is most visible. Marconia, for all its newness, is not modern: at Pisticci Scalo there is an outpost of Italy's economic miracle. The railway stations, of course, have been Pisticci's main point of contact with the rest of Italy since they were built in the nineteenth century. Each has a dusty forecourt and living quarters for the railway staff and their families in brown stucco buildings; there are two single-track railways, one of them narrow gauge. This is state enterprise of the old sort, peaceful, paternalistic and slow-moving. Morse code was used as a means of passing messages along the line until 1959. By contrast, there is a new petrochemical factory on the north side of the railway, built by the *Azienda Nazionale Idrogenerazione Carburi* (National Fuel Hydrogenation Corporation – ANIC). It is a striking sight, with gleaming aluminium retorts, pumping stations, synthetic fibre mills, and modern offices. This factory has a teleprinter and an airstrip. Near it are tall modern blocks, part of an abandoned plan to build an industrial village, inhabited by employees of the factory whose homes are not in Pisticci. *Azienda Generale Italiana Petroli* (Italian General Petroleum Corporation – AGIP)[5] has built a motel and a restaurant. Other services established at the Scalo are two other restaurants, two motor mechanics, a baker, a tobacconist, a butcher, a post office, and three petrol stations, two of them owned by AGIP. These,

[5] Both ANIC and AGIP are subsidiaries of the *Ente Nazionale Idrocarburi* (National Hydrocarburants Board – ENI), the petroleum corporation in which the state has a majority shareholding.

together with a small factory for making *Amaro Lucano*, the local digestive, are local enterprises which have moved to the valley to serve the industry, or to take advantage of the infrastructure created for it.[6]

II

Pisticci town is 1,200 feet above sea level, built on a saddleback hill. There are two big squares: Piazza Umberto, with the *Municipio* and the field offices of ministries and of associations, is the place where people sit and wait for work, where political meetings are held, where the aged men wait and watch. Piazza Repubblica is duller – a magnate's palace on one side, a church on another, private houses and a few shops. They are joined by Corso Margherita di Savoia, the main street with specialist shops; it is the place where people promenade in the evenings, and on Sundays at midday.

The rest of the town is private: the small shops – selling groceries, string, floor-cloths, socks for children – serve very specific localities. There are artisans' *botteghe* which serve perhaps rather wider areas. But if you walk in one of the narrow streets off the centre, between the whitewashed houses, nobody knows what you could be there for: if it is not your base area you arouse speculation. In the centre you do not: a Pisticcese expects to see Pisticcesi he does not know in the centre, but not in his own neighbourhood. The 5,500 houses in the town are cramped into an area 1,000 yards long, 300 yards at its widest: some of the houses are perched on the edge of the precipitous slopes which are the limits of the built-up area.

Houses are for the most part windowless; light and air come in through the door and through an unglazed grille near the roof. Traditionally, there is one room, with a recess at the back which is used for storage, keeping animals, or as a kitchen. Floors used to be built of brick, but synthetic marble tiles are mostly used

[6] About infrastructure: it should be observed that the Scalo is on the route to be traced by the Basentana, a four-lane highway which will join the Taranto-Reggio Calabria highway to the Salerno-Reggio Calabria branch of the Auto-strada del Sole. This is an ill-fated road, begun in the '50s and still unfinished. In 1967 there were 13 km. from Pisticci Scalo to Ferrandina Scalo, but it had no 'outlet' except the old provincial roads to the towns on the hills, and a narrow unmetalled track which leads further up the valley. Government maps of developing areas, however, show the road as completed. See, e.g., *The Economist*, 28 March 1964, p. 1244.

today, and are greatly aspired to. It is also considered an improvement to divide one large room into two with a wall perhaps six
feet high, two or three yards from the door. The family then lives
in the narrow front space during the day and sleeps in the back
room at night. The front space is also used for display. Refrigera-
tors, television sets, elegant modern gas-cookers with their glass-
fronted ovens filled with the family's most ornate best coffee-cups
(Pisticci cooking does not require an oven) are common sights,
together with photographs, mirrors, pictures of saints, commercial
calendars and a glittering table lamp or chandelier – all these and
the common cooking pots, tables and chairs may be crammed
into the front room of a well-to-do, peasant or labouring family.

In the back are the sleeping quarters: the matrimonial bed
covered with a day-time bedspread, and smaller beds or couches
for children, a dressing-table, two bedside tables, at least one
wardrobe (often of shining walnut, with mirrors which have
designs cut into them), and the large chest in which linen is kept.
As a general rule pictures of the Virgin Mary and other saints are
put above the head of the bed, and more profane nymphs, sylphs
and swains, lit by the moon, hang on the wall facing the bed.

For most Pisticcesi such a house, furnished in such a way,
represents that combination of prestige and comfort which a
house ought to have. Many of the poorer houses still have brick
floors, and simple furniture. The bedspread may be plain white
rather than ornately embroidered cloth; there is often no television. Essential furnishings, however, are the matrimonial bed,
the chest for linen, a wardrobe, a table and six or more chairs.
Middle-class comfort also sometimes sacrifices the more prestigious furnishings, and in these houses a visitor is given a relaxed
and homely hospitality. Families which live in this way have
unequivocal claims to high standing in the community. When a
family claims prestige through its furniture, extreme discomfort
may be forced on the visitor because every effort is made to be
impressive: the furniture is gilded, the upholstery is tapestry – often
as not covered with a sheet of plastic to preserve these conspicuous
objects from being worn out by those who use them. In these
households such domestic comfort as there may be is banished
behind the cream and gold double doors.

Of the 3,942 houses reported as occupied in the census of 1961,
one-third had running drinkable water in the house, and rather

less than 300 had baths or showers; 800 had no lavatory, 700 had no electricity. These figures refer to all the territory of the administrative district of Pisticci. A survey in 1961 classified one-quarter of the town houses as 'very unhealthy', rather more than a third as 'unhealthy', and the rest as either adequate or improvable in roughly equal proportions.[7]

Each group of streets has at least one small shop which sells mostly food, but also articles of general use such as knitting wool or floor-cloths, clothes-pegs or cheap clothes for children. The more specialized shops, selling domestic electrical equipment, ready-made suits or state monopoly goods (stamps, tobacco, salt, matches) are in the centre of the town on the Corso Margherita. So, too, is the daily fruit, vegetable and fish market. The monthly market stretches for half a mile along the streets at the eastern entrance to the town; twice a year there is a cattle market on the football field.

Pisticci is the centre for the various health services. There is one doctor who visits a surgery at Marconia, but all the rest, and the dentists and chemists, are in town. The municipal midwife and doctor, who are appointed to give their services free to the poor, are also based in the town; the doctor, in fact, rarely has time to leave it.

Each of the seven policing organizations has its offices in the town.[8] The field offices of various ministries are in Pisticci: the Cadastral Registers (Ministry of Finance) for the neighbouring communes, Montalbano, Craco, Bernalda and for Pisticci itself are kept there; the local magistrate (Ministry of Justice) hears cases also from Montalbano, Bernalda and Policoro; the representatives of the Agricultural Inspectorate (Ministry of Agriculture) have powers in Pisticci and Craco; the Ufficio del Registro (Ministry of Justice) registers all legal contracts and documents for Pisticci, Craco, Montalbano, Policoro and Bernalda. The surgery and records of INAM (para-statal, but formally controlled by the Ministry of Health), the health insurance institute for industrial workers, also serves the neighbouring communes.

Pisticci is an educational centre. The most popular secondary

[7] Unpublished material, from Prof. G. Marselli of Portici.
[8] These are: the town (*urbana*) and country (*rurale*) police, employed by the Municipio; and the *Carabinieri* (army); the *Stradale* (traffic police); the *Finanza* (tax and excise); the *Forestale* (forestry); and the *Pubblica Sicurrezza* (civil).

B

schools are in the town; and the head of Pisticci elementary school is responsible for running the schools at Marconia and Craco – he deploys teachers. There is a *ginnasio-liceo* (upper secondary school) taking children from fifteen to university entrance. There is an upper secondary technical (agricultural) school at Marconia, and a lower secondary agricultural school at S. Teodoro-Casinello, a rural centre in the Agrarian Reform zone. Single-class, elementary schools in the country are staffed by the youngest teachers.

Pisticcesi regard these rural or hamlet schools as inadequate; and I have no doubt that in many cases they are right. And although agriculture is the chief source of livelihood for most of them, they do not want to send their children to farm-training schools at any level. Agriculture has lost its prestige, and most people want their children to be industrial workers, if not to follow clerkly or professional occupations. Since there are no industrial training centres in Pisticci, parents and children prefer the more general education given in the ordinary secondary schools in the town. They recognize that the elementary schools there have better amenities and more experienced teachers than in the country. Thus, even though the schools are more dispersed than other services, there is a general tendency for parents to send their children to town schools, and to arrange that period of their lives to fit in with the children's needs.

The territory of Pisticci is a unit of local government, which is in the hands of the town junta (*Giunta*), an executive committee elected by the town council (*Consiglio Comunale*), which is itself elected by universal adult suffrage. The junta controls the *Municipio* – the permanent offices for administration of the town; it is headed by the mayor, and is responsible to the council. The mayor himself is an unwilling Janus, responsible to the provincial prefect (a field-officer of the Ministry of the Interior) as well as to the electors and their other representatives. The council assembly hall, and the offices which are the physical *locus* of the *Municipio*, are in the Piazza Repubblica, next to the church. It is here that Pisticcesi go for their business with local government – for licences, payment of certain taxes, complaints, and audiences with the mayor. The Public Works department runs its relief and maintenance work from this building. A sub-office at Marconia issues some certificates, and receives and registers documents for those who live in that area.

Thus the main day-to-day contact between the state and the individual Pisticcese is concentrated, in one way or another, in the town: not only tax offices and legal departments, but welfare services, agricultural services, education and local administration.

So, too, are the formal voluntary associations which either have delegated powers or attempt by collective effort to influence the administration. In or near the *Municipio* building are the offices of all the main associations: the Hunters (which issues gun licences), the Motor Club (ACI – it issues Driving and Road Fund Licences), the Communist Alliance of Peasants, the socialist-communist Chamber of Labour (CGIL), the Catholic Direct Cultivators. The Demochristian, Socialist, Social Democrat and Liberal parties have their offices and club-rooms here, as do the catholic workers' syndicate (CISL), the Catholic Association of Italian Workers (ACLI), the War Invalids, the War Veterans, the *Circolo Metapontino*, a club for professional men and landowners, and the *Circolo Zanotti-Bianco*, a cultural club for young teachers and students. I list these various associations to show to what extent formal inter-action outside the family is concentrated in the town, and in a relatively small area within the town. None of these associations has offices or sub-offices in the country or villages.

Religious associations of all sorts are divided by parishes and meet in the four parish churches or in rooms near to them (one of these is at Marconia). Except for the more explicitly political associations, most of these clubs have provision for entertainment – cards or miniature billiards or television. Indeed, since pin-tables were banned by law, it has been possible to play them only in the rooms of the Catholic youth association.

III

No one can really be independent of the town. People go there for all the foreseeable events which may require the assistance of a doctor, chemist, priest, party secretary or a powerful friend. Very few people have any general secondary education after the age of fourteen, but those who do prefer to attend town schools. Commercial activities are concentrated there, as are the important political and administrative relations of groups and individuals.

A man who lives permanently in the country is considered less civilized than one who lives in the town, or even one who leaves the town at four or five each morning and returns late most evenings. Life in the town is said to be more civilized (*cchiu civil'*) and the people who do not live there are *cafun*, oafs devoted to their land by choice or necessity; deprived of friends and influence, they are people of no account – *gient' da poco cunt', cca nun se calculesc* – whose interests are not 'calculated'.[9]

The town is the arena for that display in which a man gains the approval of his fellows. The main events of his life are occasions of display, and the rituals are performed in town. His baptism, first communion, marriage, mournings and funeral are all public events, and are observed and commented on there as they can never be in the country.

'A girl who puts on her finery', I was told, 'doesn't walk up and down country lanes.' The evening *passeggiata* up and down the *Corso* is the main occasion of display for well-to-do girls, as well as the only opportunity for talking to young men not of their family's circle. The most tender and intimate conversations of a girl's unmarried life may well be carried on in the public eye, walking where all can be seen but only snatches of conversation overheard. So, of course, are the preliminary glances, smiles, rebuffs and tense interchanges of a few words with an air of ostentatious indifference. Girls who are not well-to-do have other opportunities to meet or be observed by young men, but these, too, arise only in town. They go to the public fountains for water, they go on errands to shops, or they 'visit their kin'. The festival of the town's patron saint, as well as the Sunday masses and the innumerable lesser processions and services and weddings and carnival parties, are other occasions on which unmarried couples can meet safely under the public eye.

It would be difficult to over-emphasize the importance which Pisticcesi attach to the town, and more particularly to the social sanctions which are exercised within it. People who live in the country are morally suspect just because their every movement cannot be known, because it is not known what they eat, what they buy, whom they talk to. It is maintained that women, however demure they may be in town, become lascivious and un-inhibited when they leave the built-up area.

The town, then, is the home of the Pisticcese, where he conducts most of his social activity. It is there that his honour is recognized,

[9] See Davis (1969, b) for a fuller account.

and there he feels his women and his honour to be most secure. The town is distinguished in dialect, where the words *u paes'* connote home, security and civilization. Although the Italian word *campagna* (country) is now increasingly used in dialect, the ancient word for country is not a noun at all, but the preposition *for'* – outside.

IV

Economic roles are not highly specialized in Pisticci. Peasant farmers perform most of the different tasks on their farms, and rarely specialize in the production of particular crops. They produce most of their own food, as well as marketable surpluses; and most of them can turn their hand to odd jobs such as making fences or baskets, saddlery or medicating sick horses or mules. I knew one man who not only built a new storey onto his town house, but made the bricks for it first; he was praised as particularly skilful and versatile. Some men are specialist pruners or pig-killers or well-diggers (these are often water-diviners as well); but they do not do these things only, and when they do exercise their special skills for another person it is usually as a favour for a friend. Pisticcesi do not like to work for others, or to depend on them, if they can help it, and most jobs of this sort, however well paid, are called favours. When work is slack, many peasants try to find casual jobs to fill their time; and a man with too little land to employ him fully usually has a set of friends who take on him and his mule at low wages. It is becoming increasingly common for richer peasants to buy tractors; and when their own land has been ploughed, harrowed, harvested, they send a son out to work some other man's. The marginal costs are minimal, the extra income is welcomed; but this is not specialized labour.[10]

The supply of labour is well in excess of demand. This is best shown by a rough calculation. First, the average demand for labour per hectare of farm land is now 26·5 man-days per annum,[11] and there are 21,414 hectares of farm land in Pisticci. The total demand is therefore roughly 568,000 man-days per annum. Secondly, the working year is 294 days (allowing 70 public holidays); and there are 1,970 men and 1,754 women given in the

[10] See below, Chapter 6, for a fuller discussion of peasants' work.
[11] Cupo (1965), App., Tab. 41.

1961 census as earning the main part of their living from agriculture, either as farmers or labourers. A woman's working day is
generally calculated at three-quarters of a man-day. The labour
supply is therefore roughly 966,000 man-days; and there is a
surplus of supply over demand of about 398,000 man-days. This
surplus is 41 per cent of the supply of labour, and if it were equally
distributed among the work force each person would have work
for about 172 days out of 294.

Although about 55 per cent of the farm land in Pisticci is held
in medium to large holdings (of fifty hectares or more), these
provide relatively few jobs. They are highly mechanized farms
and require skilled men. In 1965 only eight landowners had a
permanent labour force of more than ten men, and they did not
substantially increase the number of their employees in the busy
seasons. Another twelve employed between five and ten men
permanently; and twenty employed fewer than five. In effect, the
whole of the rest of the agricultural labour force earns its living
either from its own land, from casual labour on small farms, or in
some other sort of work. At a generous estimate of the amount
of labour absorbed in permanent jobs with landowners, there is
still a surplus of about 300,000 agricultural man-days per annum
in Pisticci.

So, as well as being unspecialized in their agriculture, many
farmers are perforce part-time farmers. Indeed, in spite of the
clear-cut nature of the census figures it is closer to reality to talk
of a continuum between industrial and agricultural employment.
Relatively few people are employed fully in either, and most
people combine so-called industrial jobs and farm work in varying
proportions.

One reason for this is that job-security in local industry hardly
exists; and underemployment is as common in, say, the building
trade as it is in farming. Unfortunately, no dramatic confrontation
of the figures for the supply and demand of labour is possible for
this sector of the local economy; but the instability of so-called
industrial employment can be illustrated in other ways. In 1961
some 1,750 people were recorded as gaining their living from
non-agricultural work other than commerce and administration.
Of these, craftsmen and artizans were 37 per cent (649) and 63 per
cent (1,096) were builders. The electricity company employed the
other five men. Only eight local firms employed more than five

men. One of these was the municipal road-sweeping and garbage collection unit (nineteen men), and two others were building firms. When the petrochemical factory was built in the valley below the town, between 1961 and 1965, no local contractors could cope with the complexity of the task, nor with the extensive infrastructure – roads and water supply and conservancy – built to service it. The outside contractors who did the work brought a large part of their labour force with them; and by 1965 when it was nearing completion there were only fifty-six Pisticcesi employed – and forty-four of these were laid off in the late spring of that year. The factory itself took on a labour force of young men and women with high educational qualifications, and the labour market for people over thirty returned more or less to its former state.

There are two other major sources of employment. The first is the Public Works Office of the commune, which employs unskilled labour to maintain public buildings and roads. This is essentially relief work, and it is financed partly from the commune's funds, partly from the Poor Relief Board (*Ente Communale di Assistenza*), and mostly by grants from funds administered by the provincial prefect. This work is scarce, and tends to be concentrated in the winter months, with bursts of activity shortly before the major festivals at Easter and in August. Since the jobs are relief jobs the commune does not pay insurance contributions; hence the work done is not taken into account when unemployment benefit is calculated (this is done on the basis of the number of weeks worked in the two years preceding the application for benefit). In 1962, 308 men worked for 4,200 days – an average of 13·5 days each. In 1963, 332 men worked for 5,800 days – an average of 17·5 days each. The normal pay was 500 lire a day. The average pay for a casual farm labourer at that time was 1,700–2,000 lire.

The second source of work is the Regional Forestry Commission (*Ispettorato Regionale delle Foreste*), which is carrying out reafforestation work on the hills round Pisticci. The men work in gangs of about a dozen, digging terraces, building fences, planting conifers. As many as thirty people may have a job at any one time, but this work is not permanent either. The Commission is a department of the Ministry of Agriculture and is obliged by law to make graduated redundancy payments to anyone who is

laid off after more than sixty days' work. No labourers in Pisticci
are taken on for more than fifty-four days at a stretch, and most
of them work for rather shorter periods. Wages are between
1,750 and 2,000 lire a day, and although this work is obviously
offered as relief, insurance contributions are paid, and the days
worked count towards unemployment benefit. However, a man
who has been laid off, even for a technicality, must re-apply for
work through the labour exchange, which has an order of priority
for the unemployed: those who have most need come first, and
the more recent the employment the less the need. People who
have no other insurance contributions can expect to work for
two or three periods of ten to fourteen days each year; and this is
not sufficient to qualify for unemployment benefit.

So much for the opportunities of work for an industrial
labourer. It is clear that, apart from the brief interlude from
1961–5, unemployment and underemployment are as common in
this sector of the economy as in agriculture. This is the main
reason there are so many part-time farmers and part-time builders
in Pisticci: why economic roles are so unspecific.

The structure of local 'industries' is very much like that of
peasant farms. This is what would be expected of the 649 artizans
in the town, but it is true also of the building industry. Of the
thirty or so registered employers of building labourers, only ten
operated more or less continuously in 1965, and of these ten,
eight employed fewer than five men. The other twenty were
master-builders (*maestri*) who took on a bricklayer and a lad
whenever there was work to do; as an official of the labour
exchange remarked, they were entrepreneurs without enterprises
(*imprenditori senza impresa*). Capital equipment, too, was minimal;
spades, ladders, trowels and a few sticks of scaffolding are all that
are needed to build a traditional Pisticci house. Five of the ten
continuous operators had more sophisticated equipment: mecha-
nical lifting apparatus, or a concrete mixer, or both. But one of
these operators never used his equipment. A master-builder
reckons to turn his hand to anything that may be necessary, from
bricklaying to roofing, from carpentry to the plumbing which
is increasingly demanded. The title master-builder is won by
acclamation, not in any formal way; a master-builder is a man
who has been invited to build a house for somebody.

In the nature of the Pisticcese urban design there are no sealed-off building sites. Cement is mixed in the street; the men rub shoulders with passers-by; the rate of work is controlled by fellow-workers and by persistent pressure from the owner of the new house, who is often a kinsman of the builders, or who chose the *maestro* on grounds of some special relationship. There is often no incentive to finish one job to get on to the next, since very often there is no next job. There may be a complete halt to work as the bricklayer goes off to bring in his harvest; one employer regularly employed fewer men than he could 'because you stretch out the work that way'.

Local industry, therefore, is not structured like a modern industrial firm. There is little specialization and little capital; only a few firms are in continuous existence; work, while it lasts, stops and starts, fluctuates with the seasons and is not done in a special environment.

With work so scarce, and competition for it so keen, the few who have permanent jobs are in a very weak contractual position against their employers. This applies as much to the farm workers as to others. Bus drivers in Pisticci, for instance, have never joined one of the fairly frequent national strikes; nor do they negotiate collectively for improvements in their working conditions. They work seventeen hours a day for two weeks, and fifteen for the next two, and so on. With two Sundays free out of four, and one and three-quarter hours for meals each day, they work an average of 110 hours a week. They not only drive the buses but clean them inside and out, and help with minor repairs. The monthly wage in 1965 was 75,000 lire; and they had two weeks of unpaid holiday a year, from which were deducted any days lost through illness.

Although the bus drivers are hard worked, and have little chance of improving their conditions, they are relatively well paid; and the chief attraction of their job is its security. With a steady income, however small, they are in an envied position. The relationship between employer and permanent employee is not an impersonal contractual one; rather than an impersonal figure the employer is *padrone*, a patron, possibly a friend. The rules are applied with some leeway, and always in friendly language, jokingly. It is assumed that to get a job at all the prospective employee must have some special relationship with his employer; certainly other considerations than his specific skills are taken into account. The employer's influence can be used to secure a job for a son, or an easy passage through his military service, and so on. The relationship is many-stranded, not managerial and single-stranded; the worker is not simply a worker, contracted to supply a certain amount of work, but a client.

In Pisticci every man can take on most of the jobs associated with his main occupation, and main occupations are rarely sole occupations. The boundary between farmer and farm labourer, or between agricultural worker and building labourer, is vague and frequently crossed. Farmers, farm labourers and building labourers are the majority of the active population, but the same pattern of underemployment and part-time working applies to white collar jobs: school teacher, lawyer, clerk and shopkeeper.[12] Workers do not have precisely demarcated skills, and work itself is not divided and subdivided into specialist tasks. Perhaps a permissive cause of this is that some of the main sources of employment are of the same sort; peasant farms and building firms are equally characterized by low capitalization, seasonality and, to a lesser degree, the absence of an isolated workplace.

This sketch of the economy has shown that economic roles are not so specific as might be expected in a developing community: not only do peasant farmers – like most peasant farmers – produce a lot of their own working capital, but they are likely to become labourers of one sort or another for a week or two when they have the chance. There has been expansion in the 'industrial' sector, but this has not been accompanied by specialization.

V

I do not intend to present a sociological analysis of politics in Pisticci. The following account is anecdotal, but it may serve to support two very general points about the specificity, or lack of it, of political roles. The first is that since many tasks which are not specifically political are devolved on the political parties,[13] many more people are involved in one way or another in political activity than are in, say, Britain, where politics tends to be very much a specialist's affair. And the second is that the principles of

[12] Davis, (1969, b).
[13] The (Catholic) Direct Cultivators' Guild in Pisticci, for example, runs the peasants' pension scheme for *all* peasants. Some details of administrative tasks devolved on other associations are given above, p. 9.

recruitment to parties are not necessarily views about which party ought to govern, or which ideology is fairer or more just than another, but may include kinship, friendship or gratitude for favours received.

The minutes of the communal council and the administrative junta, and the records in the Potenza archives of correspondence between the prefects and mayors and the heads of the local police force, show fairly clearly that up to 1921 the politics of the town were dominated by a few rich families who controlled a minority electorate.[14] Political and economic dominance went together. In 1912, when the suffrage was extended to all adult males, there began a brief interlude – until 1913 when the elected council was suspended more or less for the war's duration – of 'popular' government under a republican administration, led by a lawyer, *Avvocato* Alessandro Bruni. After the war Bruni again controlled the council but he was forcibly thrown out, and then banished by police order in 1923 when the administration was reformed under a fascist party *Podestà*. By and large, the administration was again dominated by the rich landowners. After the second war the republican party again had a brief flowering, but was ousted by the communists, who took over most of the republican following. Since then the administration has been controlled in turn by the demochristian (government) party and the vigorous communist party; between 1963 and 1968 the town had a communist ex-mayor in Parliament. Until 1965 the demochristians were allied with the *Movimento Sociale* (the neo-fascist party) which had one or two representatives on the council; in 1965 they formed a centre-left coalition with the socialists, gaining control from the alliance of communists and proletarian socialists. The variety of political experience in Pisticci should not be underestimated.

A man who chooses to participate in formal political activity is assumed to do so in order to further the interests of his family, and people point to the ways in which a successful politician controls public assistance, relief work, public contracts and the investigations of the police into allegations of misappropriation of funds. If there is a conceivable interest, that is the explanation of political activism. But people can and do choose to operate the political system on specific political criteria; they may choose to

[14] Some details of these families, in relation to their manipulation of the distribution of the demesne, are given below in Chapter 6.

give their vote, or to join a party, because they share the aims and
ideology of the party rather than because they are kinsmen of the
man who is standing for election; they may refuse as a matter of
principle to take gifts to elected men who control funds, because
they claim they have a democratic, citizen's right to them. And
my impression, gained, however, in only a year, is that more people
are choosing to act in this way than formerly.

Nevertheless, important politicians are supported by their kin,
and on occasion mobilize them to swing an election or organize a
vote. All the recognized kin of the secretary of the *Alleanza
Contadini* (Peasants' Alliance, a communist party organization)
are ostensible PCI voters, and go to the secretary for aid in their
business with the bureaucracy. Some of these men, however,
were not CP voters in fact and in private expressed disquiet at
the proceedings of their kinsman; but they attended his meetings
and regarded him as a useful kinsman to have. A large number of
the members of the alliance were not his kinsmen and chose to go
to him apparently because he and his organization were more
efficient at securing their rights than rival organizations.

In the administrative elections of 1964 (when communal and
provincial councillors were elected) there was a quarrel within
one of the major parties. A town councillor put himself forward
for election to the provincial council against the opposition of the
leaders of the local party, who wished to reserve the provincial
arena for themselves. The man was elected to the town council,
but got less than 100 votes to the provincial council. In the
quarrel, which was conducted at mass meetings in the town
square, the defeated councillor claimed that the leaders had
mobilized their kin: their wives and sisters, he said, had gone to
their neighbours and their kin to tell them to vote for him in the
town elections, and not in the provincial elections: he made
various other charges, but this one, which was a striking admission
of the power of the party leaders, was never denied. The point of
this anecdote, which was unique in my experience though I was
told of other similar histories, is to suggest that political ties are
not sufficient in themselves to manipulate the political system and
that a party in Pisticci consists of people 'united' not simply by a
common political creed, but also by other common interests and
obligations; party membership tends to represent a concentration
of relationships of diverse sorts, rather than a single political one.

The other main source of support for parties appears to be the success of their local officers in conducting business with the state bureaucracies on behalf of the members. The main political parties and the labour syndicates have statutory powers to help their members obtain documents and welfare benefits from the state; and the state pays a fee to the organization for each completed bureaucratic exercise, graduated according to the difficulty of the task – more for a pension than for a passport, say. The fees are paid to appropriate institutes set up by the parties, and the local party officers are the agents for the institutes – which are normally dubbed 'for Social Assistance' (*di assistenza sociale*). The fees are then distributed down the organizational hierarchy, and part of them ends as a supplement to the salaries of the local agents.

Apart from the direct and legitimate financial interest which parties and syndicates have in attracting members, a large member-ship is of more general importance to the local officers. The secretary of a large local branch has more say and more influence with his superiors in the hierarchy of the association than the secretary of a small one. There is, therefore, keen competition for members not only between ideologically opposed associations, but also between different associations of the same political colour.

Shortly before the petrochemical factory began to take on staff the Pisticci branch of CISL, the Catholic Workers' Syndicate, had four members, no office and no secretary. A room was hired, and a secretary appointed by the provincial office. The new secretary found that, as he put it, 'the functions of the labour syndicate' had been taken over in part by the Direct Cultivators (the Catholic Peasants' Guild), in part by the Mothers' and Children's Institute (ONMI – not an association, but a welfare bureaucracy, the secretary of which was an influential local citizen) and by the ACLI, a recreational and morally educative association for Catholic working men. The new secretary of CISL fought at first, not the communist-dominated Chamber of Labour, but other pro-government organizations which had taken over his potential membership. When he had finally obtained a directive from higher authorities to the secretary of the local Direct Cultivators that industrial workers should be referred to CISL, and refused admis-sion to the peasants' associations, membership of CISL increased from four to nearly a thousand in three months. Only then did he set about winning members from the communist organization.

To do this, he decided, he had to show that he could conduct business with the provincial welfare boards more efficiently than the communists. He arranged that a member of the provincial committee which examines requests for sickness benefits and pensions should come to the Catholic Syndicate's office in Pisticci twice a week to consult with the members about their problems, and to help them fill in the complicated forms of application.

The task of helping people in this way is important: the bureaucracy is a complex one, and does not readily give what citizens expect from it.[15] The main parties do not insist that the people they help should be party-members; but a man who uses the party organization successfully may well join out of gratitude. The secretary of the Communist Peasants' Alliance was reputed to be the best man to go to, and he claimed to deal with about 4,000 applications a year. This was probably an exaggeration[16]; but secretaries of other organizations appeared to do rather less than he did.

My own impression was that the communist secretary was a virtuoso intermediary, able to lead exciting guerrilla attacks on the bureaucracies. His counterparts in other organizations had greater and more constant access to power, but were themselves bogged down in the red-tape of their own organizations. On one occasion I visited the farm of a peasant, known for the forthrightness and wisdom of his knot-cutting counsel, together with the secretary and his father-in-law, to pick fruit for the secretary's own use. The *quid pro quo* for this favour was that the secretary would undertake a particularly complex and tricky negotiation with the authorities. Although the peasant had been separated from his wife for more than twenty years, the secretary would try to get his pension increased by 2,550 lire a month on the ground that his wife was still officially dependent.

Granted that some people in Pisticci and, in my opinion, increasing numbers of them, make their political choices on what we should accept as strictly political criteria, I think these anecdotes have served to show that politics in Pisticci are not specific, and in two ways: the institutions themselves serve a variety of purposes; and people join them for a variety of reasons.

---

[15] See the interesting account by Pizzorno of the difference between Roman-law based administration and others (1968).
[16] See, for example, the files in Plate 1*a*.

VI

There are some people in Pisticci who have never been to the town's railway stations and have seen them only from the hills. There are some who cannot read or write; some people believe that rockets are taken to the Vatican for a blessing before being launched to the planets; some people thought that English was another dialect, and indeed referred to Italian itself as Florentine. The catalogue of quaint beliefs, of ignorance, poverty and sheer beastliness in relations with other people could be a long one. And it is true that Pisticcesi are, on the average, poorer, less well nourished, less well educated, and have fewer opportunities than industrial workers or even other peasants in other parts of Italy.

But it is a mistake to think of Pisticcesi as wholly cut off from the sort of life we know. There are peasants – one or two – who go abroad for holidays. And it is not the expense which makes these trips remarkable, but the awareness of possibility and the freedom of choice. All men have gained some knowledge of other parts of Italy during their military service. Most of the older men have fought in one or other of the two world wars, or in the Ethiopian war, or taken part in the imperial colonization of North Africa. Pisticcesi have a wider political experience – of different sorts of government – than any native of the United Kingdom; and they have, today, a much wider range of political choice than we do: a Pisticcese can vote at either extreme of the political spectrum without being eccentric. Because they have worked as labour migrants, many young men have a wider knowledge of working conditions in factories in Western Europe than their counterparts in Britain. Pisticcesi are dependent for many things on the central government, and the leaders of the community enter into sophisticated negotiations to get funds for roads, housing projects, for a hospital, for bringing the electricity supply to farmhouses.

None of this alters the fact that they are poor, that they are forced to go to Germany as labour migrants, that their politics are 'undemocratic'. But the lack of a modern political and economic structure does not have as its consequence a limited and homogeneous life-content. A Pisticcese can choose between several ways of doing things, and he will not be labelled eccentric or deviant as the result of his choice.

2

Marriage

I

In Pisticci married status is equivalent to adult status. A man does not generally make contracts, rent land, take part in politics, initiate patron–client relationships until he is the head of a family. This usually means he must be married, though a few men are official heads of their families because their fathers have died. A woman also becomes an adult only when she marries. Until then she is not only under the tutelage of her father and brothers – a tutelage which will be replaced by that of her husband – but she has little independence in domestic affairs; a good daughter works hard for her mother and takes few decisions. When she marries a woman gains not only domestic independence but access, in her own right, to the group of the women of the neighbourhood. This is her chief source of power, since it is the members of this group who are the primary agents of informal social control.

Many discussions of honour in Mediterranean societies concentrate on the formal qualities of honourable men or women; or they consider the conceptual system which various sorts of honour combine to make. Here I shall emphasize the sources of honour.[1] In part these are hereditary: a family has a good name, a good reputation which is known and repeated as a part of the justification for the position which it holds in the ranking system of the neighbourhood. Inherited honour, however, can always be modified. Although an inherited reputation often is a self-fulfilling prophecy, it is not an absolute determinant.

Achieved honour comes from being a good adult – a good husband or a good wife. The main source of honour is in domestic life, in the way a man or woman behaves as a spouse and parent. The connection between domestic reputation and the wider societal honour in a society such as Pisticci appears to be that where rules of behaviour are sanctioned informally, political

[1] For a fuller discussion, see Davis (1969, a).

activity and patron–client relations depend on trust: a man must have a reputation, and he must have something at stake when he enters the political arena.

I have argued elsewhere[2] that patron–client relationships are like gift–exchange relationships except that, tactically, a man may in some circumstances fail to return a 'gift' and still not lose reputation; he has been *furbo*, cunning and crafty, and has outwitted his partner. The quality *furbo* – foreseeing all the possibilities, having the strength of purpose to ignore the moral and affective claims of a partner and to take advantage of his gullible, or trusting, or slow-witted nature – is much admired in Pisticci and is inculcated at an early age. A cleverly naughty child may be punished if he annoys his parents, but after they have calmed down they will praise him as *furbo*; if he annoys other people, the parents will comfort him by telling him how *furbo* he has been. I was once treated to a performance of the following catechism by an eight-year-old boy:

FATHER:	*Sei tu bravo a scuola?*	Are you clever at school?
SON:	*Si, sono bravo a scuola.*	Yes, I am.
	(this repeated three or four times)	
FATHER:	*Tu non sei bravo a scuola?*	Are you not clever at school?
SON:	*No, sono bravo a scuola.*	No, I am.
FATHER:	*Sei tu fesso?*	Are you slow-witted?
SON:	*No, non sono fesso.*	No.
FATHER:	*Tu non sei fesso?*	Are you slow-witted?
SON:	*No, non so fesso.*	No, I am not.
FATHER:	*Sei tu furbo?*	Are you quick-witted?
SON:	*Si, sono furbo.*	Yes.

The child, when he had successfully dealt with the negative questions sprung upon him, was rewarded with hugs and kisses.

In adult life men's successful enterprises are attributed to their *furbezza*, as well as to the power of their patrons. The indigenous landowners of Pisticci who rose to control the local economy between 1880 and 1940 at the expense of the absentee landowners are the local paragons of *furberia*. Pisticcesi peasants do not tell jokes or funny stories, but rather anecdotes about real people or their ancestors; these often turn on someone being *furbo*, or its opposite *fesso*. One family, for example, is nicknamed *Salat* (salty)

[2] Davis (1964).

C

because a grandfather had evaded the state monopoly of salt by growing some of his own. *'Era così fesso, l'ha piantato'* (He was so *fesso*, he planted it). The success of certain contemporary builders and shopkeepers is explained and illustrated with accounts of how *furbo* they have been. A common response to a proposal for any collaboration of an unusual kind is *'Mi prendi per fesso?'* – 'Do you take me for *fesso*?'

Honour, reputation, is the basis for trust in Pisticci, where all dealings outside the family are assumed to be governed by self-interest, with *furberie* as the chief tactical devices. A man who is known as a sound family man is to some extent calculable: he will not take risks which, if they turned out badly, would damage himself and his family. Similarly, if a man's proposals or behaviour are mystifying, the key to understanding is thought to be his self-interest rather than, say, his altruism. Sobriety, fair dealing and a calmly straightforward firmness are qualities as much admired in husbands as they are required in business or political associates. How a man behaves in his family is, therefore, the guide to his probable behaviour in the less closely controlled and supervised political and economic spheres where, since competition is so acute, most actions appear conspiratorial, and all plans are schemes. The simple fact that he has a family at all, is a married man with adult status, is a guarantee of some element of responsibility.

The incentive to marriage is thus not simply autonomy in domestic life. The married status establishes the right, for men, to full participation in political and economic life, and, for women, to play a full and independent part in the life of the neighbourhood.

A man attains adult status when he marries, and he gradually loses it as his children marry: he becomes a respected elder, though there is no term for this in Pisticci.[3] When a man's family begins to disperse, his need to make *combinazioni* of different income sources diminishes, and he retires from active political life. His property also diminishes, as, ideally, he endows his children with it when they marry. Pisticcesi say this explicitly: 'I have laboured; I had *n* children, and they are now all set up: I am retiring from life.'

Marriage then marks the entry of a person into adult life, and

[3] People who are generally respected, and have influence in their neighbourhood or in the town, were addressed as uncle or aunt (*zi, zi'*). See Davis (1972).

further marriages mark his departure from it: as his children marry he retires. Marriages are the cardinal points of the life-cycle, and marriage of his last child is the terminal point of an individual's full participation in the society.

II

Marriage is permitted between all persons who cannot commit incest, defined in Italian law as sexual intercourse between ascendants and descendants, or between a person and his spouse's ascendants or descendants, or between siblings. Otherwise people who wish to marry must obtain special permission if they are parent and child by adoption, siblings by adoption, parent's sibling and sibling's child; or if one is the widow or widower of the other's adoptive parent or child, or of the other's sibling (*Cod. Civ.* art. 87). A small fee has to be paid for the permission. There are absolute prohibitions on marriage between unrelated persons if one of them is insane, or if one of them has been convicted of the murder of the other's spouse. Widows may not marry for three hundred days after the death of their husbands; men under the age of sixteen and women under the age of fourteen may not marry, and they must obtain permission from their parents while they are under the age of twenty-one (*Cod. Civ.* arts. 84, 85, 86, 88, 90).

Marriages between persons who require special permission are rare. The commonest are with a deceased wife's sister. I knew of five such which were current in 1965, and there were probably more which I did not hear of. There were none with a deceased husband's brother, none with uncles or aunts. Marriages between cousins (about which there are no rules) are fairly common; there were a number of people married to step-siblings, and there was one, many years ago, between a man and his mother's brother's widow.[4] None of these, except cousin marriages, are of any numerical significance.

Most Pisticcesi girls hope to be married by the time they are twenty-five, and to a man older than they are but not more than thirty-five. The majority of men, in fact, are married by the time they are thirty-three. Marriages between people under eighteen are very rare, although there is some evidence that, four or five

[4] See below, Chapter 8.

generations ago, girls were married at the age of twelve or thirteen. Most people marry. At any given moment, about one-seventh of the adult population is unmarried, and in this seventh men outnumber women by 3 : 2.

These are the limits within which marriages are made. A desirable marriage – a good match – has additional qualities. Spouses should, so far as possible, be equals: marrying at the ages appropriate for their sex, they should be similar in health, virtue and wealth. White-collar workers, who are educated, sometimes add education to these three qualities: 'My daughter stayed at school till she was sixteen: she won't (= I won't let her) marry a yokel, however rich he may be.' Since parents on both sides make settlements on the new couple, the component of wealth could be varied to take account of irreducible differences in the spouses' states of health and virtue. However, I find it difficult to imagine any such negotiation. In the past, marriage negotiations were sometimes broken off because a hint had been made that one family was less honourable than the other. I sometimes witnessed conversations between neighbours about marriage settlements. The arrangements were then sometimes subtly disparaged. Someone would ask, 'Is the mother not related to Maria Draghignazzo?' An apparently standard reply to this – made by someone close to one of the intending spouses – would be that the endowment was a hectare and a half, or that the man had a good job, or that the parapherns were luxurious. I heard few of these conversations, but inquiry generally revealed that 'Maria Draghignazzo' was disreputable, and at the time I understood them to suggest that the deficiency in honour might be counterbalanced by wealth. On reflection, I am inclined to think that the reply constitutes not an admission of tainted honour, which would be implicit if the wealth were varied to meet the case, but a rebuttal: that is, it should be paraphrased not as 'Yes, and therefore they are being exceptionally generous', but rather as 'They are so wealthy and are giving so much with their child that they cannot be considered to be dishonourable.' My reasons for changing my mind about this are not, I must admit, any wealth of evidence, but a search for consistency: it seems to me that the association of poverty with immorality in Pisticci is so close that a demonstration of wealth must rebut, not excuse, an accusation of worthlessness. The reader is entitled to disagree.

The main reason for negotiation about the amount of property to be settled on the couple is that the prestige of the parents on both sides is involved in seeing that their child is well and securely set up with a house, furniture and a livelihood. The effects of this on relations within the nuclear families are discussed in the next chapter.

This chapter now follows the order of events from courtship to wedding; the final section resumes and discusses the information on property exchanges which is given from time to time.

III

Courtships tend to be brief and to the point. A man sees a girl who attracts him, asks acquaintances about her if he does not know her, and then proposes marriage to her family, although nowadays he is much more likely to approach the girl first. Mario, thirty-seven years old, anxious to find a wife after spending his youth in marrying off his sisters, told me he had proposed to three girls in the four days of the S. Rocco festival in 1965, and had been rejected each time. On the third occasion he had met a friend and told him his problem: he was looking for girls to propose to. The friend took him home to discuss the matter – an unusual gesture of intimacy which obviously moved the would-be husband, who noticed, too, how kindly he was treated by the women of the house. Together they drew up a short-list of possible brides, and went out to survey them. Mario stopped the other: 'It's useless', he said, 'to look any further. I want to marry your daughter'. 'If you wait a moment', his friend replied, 'I shall ask her.' After ten minutes, Mario was called into the house, where the girl told him she was pleased and flattered but could not accept because she was not used to the country and did not know how to do all the tasks required of a peasant's wife. Mario and his friend went off to check the list, but none of the girls was suitable, for one reason or another.

Not all courtships are as brief as this. An understanding may last for years before it is recognized as an engagement, especially if the couple are young. And a man usually waits to see if he is favoured by the girl – if she smiles at him, or if she salutes him in the street – before approaching her parents. The middle-classes, too, are open to ideas of romantic love, and form attachments at school or on the nightly promenade on the main street. Such undercover engagements are usually innocent and often definitive; and although the couple may feel that they are doing something which would be condemned, this is usually because they are doing something which is hidden: one such couple, for example, kept an ideal match under cover: they were the only children of two shopkeepers, rivals in a specialist field.

It is only when the couple can avoid observation – that is when the man has a car and the girl has no kin in the neighbourhood – that the pattern of courtship is different. Such men may be unmarried landowners, young professional men or rich shopkeepers; the girls are usually elementary schoolteachers, since in the early stages of their teaching careers such girls are often posted to provinces distant from their homes. Their relationship may not be wholly innocent, and it may not lead to marriage.

When a man proposes, a girl's parents may either reply immediately – and this is probable if he has taken care to inform her beforehand – or they may require time before coming to their decision; he will be told to come back in a week, and meanwhile they may make inquiries. Recently, peasants have come to be considered unsuitable husbands because they do not have the steady income and hours of work which a fully employed labourer or specialist worker has. Paradoxically, too, as a result of the introduction of new crops bringing the possibility of earning a decent living from the land, a peasant husband is less able to offer his wife a peaceful life in town than, say, a bricklayer. This is because most cash crops require intensive labour, so that demands are likely to be made on the wife's labour in the country: this is not only hard work but morally suspect.

Great attention is paid to a man's family as well as his occupation. If it is of low prestige, notoriously disorderly or quarrelsome, he is likely to be rejected by any family which is superior in these respects. If there is a history of mental or physical ill-health, he is likely to be rejected.

At some early point a man tells his own family that he is about to propose, or has already done so, and they too advise him and consent or object to the proposed match. The bride, like the man, should be of an appropriate age, healthy, wealthy and virtuous. A history of sexual misbehaviour in the family, especially on the female side, affects the suitability of the bride quite as much as her personal virtue. One important cause of a shouting quarrel, mentioned below, was that the girl's mother's sister was a notoriously loose woman. Again, while I was in Pisticci, a widower from a nearby town entrusted a Pisticcese, whom he took for a friend, with the task of finding out about a Pisticci woman who was thirty years old and unmarried. But the Pisticcese saw his loyalties lay closer to home: after the stranger had married the woman, he discovered what every Pisticcese knew: she was a mule (the child of a priest and his lover), and she had been conceived in adultery. The widower's complaint, part of his diatribe against false friends, was that he had been persuaded to marry into a dishonest family; and that, if the girl's mother had behaved improperly, the daughter was also likely to. Pisticcesi were delighted with the match.

Good health, too, is an important desideratum. Both men and women should be able to work and to produce healthy children. In spite of the spread of medical insurance it is still possible to be ruined economically by ill-health.

Health, wealth and virtue are the three considerations which determine the suitability of the marriage. The partners should be equally endowed with each quality, and a bride or groom who is deficient has to make do with a spouse of approximately equal disadvantages. A set of marriage contracts which I was able to examine includes that of a woman of thirty-two, a farm labourer with a dowry valued at 100,000 lire, who married a pensioned invalid six years younger than herself. As he was a non-earner he could not find a young or a rich bride; and she, being old for marriage and poor, found only an invalid. Unchaste brides may marry mentally deficient or deformed men; very poor women may marry widowers, and so on. One Pisticci man had a spastic and probably mentally subnormal daughter who, on account of this, had been raped several times; he summed up the whole ideology of equal marriage when he remarked, 'There is no man in the world unfortunate enough to marry her.'

One consequence of this emphasis on the equality of marriage partners is that marriage is not a common means of social mobility. Because of the limited opportunities for courtship, young people do not have the opportunity to form sentimental attachments which override material considerations; spouses tend to be drawn from categories of people who meet on relatively easy terms, and often enough these are kin. Courtships are usually brief, and matchmaking turns on the three basic desiderata. A man prefers a beautiful mate, but this is by no means a pre-eminent consideration.

If the proposal is accepted the couple are called *ziti*. The betrothal is celebrated by a small party at the house of the bride, to which the two families and near kin are invited. Between the betrothal and the marriage, a period which may last for years, the man visits the bride in her home under the supervision of her parents. There are also formal visits by the bride to his mother, on which she is usually accompanied by a married sister or sometimes her mother.

Negotiations about the property which the parents of the bride are to give her begin only when there is an engagement to marry. 'How much?' is said – not always jokingly – to be the first question a young man asks after his acceptance as a prospective son-in-law. Property is demanded from the man's family if he has no job. If he has no property the engagement may be prolonged until he has a job. The details of the transfer are described in a later section of the chapter.

Engagements are rarely broken; but if the girl turns out to be unsatisfactory it is commonly agreed that this can and should be done as soon as possible. A proverb which is used in this context is *Tuornete dal sulc e no da a purc'* – [If the ground is too wet for ploughing] turn back at the first furrow, don't plough half a field'. A girl may be considered unsuitable on two sorts of grounds: if she turns out to be personally unsympathetic or is a bad cook, or her health or virtue change; or if the parents fail to agree about property. My impression is that engagements were broken in the past more often because a dowry had been promised and not paid than because no promise had been made. If negotiations do break down this is likely to be because the negotiators, usually the two sets of parents, insult each other and take offence. And when this happens, it seems, on the slight evidence I have, that the children will as likely as not go ahead and get married anyway. I was never present at a discussion between two sets of parents, nor are Pisticcesi explicit about such disagreements; but on the evidence of hints and allusions it does seem warranted to suggest that disputes originate during the process of bargaining, and often concern the honour of one of the families. Most bargaining turns on a more or less shrewd appreciation of the resources at the disposal of the families concerned.

A previous engagement is no obstacle to a subsequent one. A girl is usually supervised strictly during her engagement, and will suffer minimal loss of reputation if it is broken off; a few men boast of their many betrothals. But a much-engaged man may find himself required to marry his bride by the civil ceremony as soon as the engagement seems serious. So, too, the parents of a girl whose betrothed has to go away, perhaps to Germany to earn money, may insist that he marry her at the municipal offices before he leaves. The same practice is used when one of the families has to go into mourning and therefore postpones the marriage. The purpose is to secure the union: the bride and her family can relax the search for a husband, since they are assured of their man.

Civil weddings are not public ceremonies as religious weddings are. The bride does not wear white, there is no reception, no procession and no gifts are given. One couple who married against the man's parents' wishes went secretly and by different routes to the *Municipio*, where the marriage was performed with clerks as witnesses. The civil marriage is indissoluble, and its legal consequences are that any child the wife bears is legitimate, and that either spouse can apply to the courts for matrimonial orders of various sorts. In Pisticci usage, however, it does not entitle the couple to live together, and they remain with their respective parents and are as strictly supervised as before. Only two Pisticcese couples were living together after the civil marriage: both the husbands were left-wing politicians, and they were criticized more severely than couples who lived together without going through any ceremony at all.

Civil ceremonies are sometimes used by children to persuade their parents to consent to their marriages: the parents are presented with a *fait accompli*, and put the best face on it that they can. Of the two marriages which were performed without parental consent while I was there, only one was preceded by a civil ceremony and this, whatever the intention, did not alter the parents' resolve. If the husband's parents oppose a marriage it usually takes place in the normal way, although most of his family are likely to absent themselves; a young man can leave his parental home and live with friends while such arrangements as publishing the banns are made. But for a bride the practical difficulties of marrying without leaving home are too great, and the couple usually elope.

Religious marriages establish the right of a couple to live together, to act independently of their parents, and to manage their own property; and the wedding is an occasion of great formal festivity. After the procession of the bride to the church, and the religious ceremony which is usually a nuptial mass, the couple come out of the church to the recorded sound of bells and a wedding march ringing out from loudspeakers attached to the belfry. There is then another procession round the main streets of the town. Nowadays this is a motorcade, and all the cars sound their horns: a lot of noise. The bride and groom sit together, and they are often accompanied by close kinsmen from each family. If the couple are not recognized, they may still be identified by their company.

The procession is followed by a reception. In rich families this may be a full-scale wedding feast attended by the close kin and intimate friends of both families. In the evening there is a second reception with music and dancing, which may continue to midnight or later. At both receptions the couple exchange compliments (*cumpliment*) with their guests: the guest offers a gift of money, if he has not already done so, and the spouses invite him to drink a glass of spirits, and to eat a cake. Guests are differentiated, and the more important and the more intimate drink more than once in front of the spouses, and may be given cakes or spiced sausage and bread to take home with them. The bourgeoisie avoid such 'vulgar' manifestations, though they sometimes achieve the same end by giving boxes of sugared almonds of slightly different sizes to their guests as they leave.

At the end of the evening party the couple leave for their new house, and the next morning they receive visitors, who should bring gifts of food (*presiènt*): grain, oil, sugar and other groceries, or perhaps a capon. Although the husband may go out briefly after a few days, neither spouse should take up normal activities for at least two weeks. During this period the mothers take turns to supply them with hot food; this used to continue for the whole period of seclusion, but now that gas-stoves have replaced wood fires, and cooking is less tiresome and dirty, the bride usually begins to cook for herself after a few days.

The cost of the wedding is borne by the husband's family. It may vary between a minimum of 200,000 lire and a reported maximum of 1m. lire. Most people spend probably not more than 400,000 lire.

IV

Property transactions at marriage are of three sorts: the settlements which provide the economic foundations of the new household, and are part of the process of transmission of property from one generation to the next; the gifts from the friends and kin of the spouses, which are part of a loosely organized series of gift-exchanges; and the traditional gifts made by the bride to members of the groom's family. These last are once-for-all gifts, as are those from the groom to the bride; they are not reciprocated in the same kind. The two families do not exchange property.

The bride's gifts to her future husband's family are called *u sport*, a singular noun which refers to one parcel of gifts from the bride to the groom, to his mother and to a future sibling-in-law – usually one who lives in the parental home and who has not received gifts at the marriage of any other sibling. These are the only recipients. It is said that a bride may make a present to her future father-in-law if she wishes, but this is not expected, no informant could remember one having been given, and there was some doubt whether it would be called *sport*. The gifts are always said to come from the bride, although it is usual for them to be paid for by her mother. The gifts themselves are clothes: a pair of socks, underpants, a shirt and a handkerchief for the groom, and maybe a few packets of cigarettes as well. The traditional gift to his mother is a white blouse with wide sleeves and lace at the wrists and a black shawl. The sibling should receive a dress or a shirt. These are all brought to the groom's house before the wedding, and are displayed there to any caller. The groom, who has usually made gifts to his bride on her name-day and on public festivals throughout their betrothal, now presents her with a wedding dress; these are now invariably white full-length gowns with veils of the sort commonly worn in England. All the *sport* clothes are worn at the wedding.

The wife's dowry[5] ought always to include a town house and paraphernalia such as linen. If in addition she has land it is welcome. Poor people cannot afford to give each of their daughters a house, and often give a cash sum instead.

The linen (*corredo*) which the wife brings is sometimes worth almost as much as the house. It is obviously a great investment in terms of money and labour. The important articles should be embroidered, the bedspreads fringed, the handkerchiefs initialled, and so on. It should all be unused, although it is not necessary that it should be new; a mother who has spare linen will give some to her daughters, as will a childless aunt. Women begin to collect the linen dowry for their daughters while they are still young; one had always embroidered dowry-linen except, she said, during her first pregnancy (when she was making infant clothes). The more usual age to begin is when a daughter is five or six years old and has clearly survived; but it may be as late as puberty. Great pride is taken in the linen, which is displayed in the bride's home before it is transferred to the new house. Bronzini[6] reports that the linen is carried in large uncovered baskets by young girls dressed in their best for the occasion: and he says that this is 'among the most significant and conclusive' nuptial events. But I found no trace of it either from questioning or observation; and since marriages are frequently matrilocal, I doubt that, in these cases at least, the processions were very formal.

The amount of household linen varies as a function of the number of sheets, but different informants give a different ratio. A *corredo a sei*, however, has six sheets, eighteen pillows, three blankets, and so on. A *corredo a tre* (a minimal one, that is) has half the number of each item. The following is a *corredo* more or less 'at ten', brought to her marriage by the only daughter of a

---

[5] It is convenient to continue as I have done so far, and to call the property the wife brings to her marriage her dowry: *dote* is the Pisticci word for this. But dowry (*dote*) is also a legal term meaning property subject to special limitations on its use. In the first, non-legal, sense it is property donated from a particular source. In the second, legal, sense the source is irrelevant; what are designated are the particular rights and duties attached to the property while the marriage lasts. I shall continue to use the word in its non-legal sense, but it should be remembered that it is possible for a Pisticcese to say both that he wants more dowry (property with his wife) and that he does not want it to be dowry (tied by law) at all. The legal sense of dowry is explained more fully in Appendix III.

[6] (1964), pp. 284–7.

landowning peasant. It is considered no bigger than one would expect from such parents.

10 sheets
1 eiderdown or warm blanket
1 woollen blanket
6 picqué blankets
30 pillows and pillow-slips (6 for each of 5 beds: 2 night pillows and 4 day [i.e. display] pillows)
2 straw mattresses
2 wool mattresses
2 sponge-rubber mattresses
5 bedspreads
10 towels
10 dishcloths
12 white napkins, 1st quality
12 coloured napkins, 1st quality
6 white napkins, 2nd quality
6 coloured napkins, 2nd quality
2 large tablecloths for 12 places
2 smaller tablecloths for 6 places
10 napkins for bread-making
10 dresses
10 *parull'* – i.e. 10 pairs underpants, 10 petticoats, and 10 night vests
4 pairs shoes
1 summer coat
1 black summer coat
1 winter coat
30 handkerchiefs *circa*

An American anthropologist[7] has said that the collection of linen is one of the few forms of legitimate saving because it does not attract the evil eye, as other forms of accumulated wealth may do. This may, of course, be true of other towns, but it does not seem to be the case in Pisticci. Evil eye (*malocchio*) is conceived as the instrument and, by synecdoche, the consequence of envy which anyone may have in his heart. Wherever there is envy the evil eye may be at work. There is nothing to suggest that sheets and handkerchiefs, among all the goods Pisticcesi prize, are uniquely immune from envy. Evil eye, moreover, is by no means a common form of explanation in Pisticci, and it would be wrong

[7] Pitkin (1961).

for us to invoke it to account for a passion so widespread as the collection of parapherns. On the other hand, the *corredo* is one of the few forms of saving open to women who do not have private bank accounts nor separate post office savings: they spend their spare cash on linen and not on lottery tickets, cards or wine. It is a mother's duty to provide her daughter with linen when she marries, and no one else shares this duty. Thus, when she collects linen a woman combines her sole traditional way of saving with the fulfilment of her unavoidable duty. For Pisticci this constitutes a sufficient explanation, and it only remains to point out that a *corredo* is a poor sort of investment; it does not carry interest, nor does it have great liquidity: it is hard to sell it, and impossible to secure a loan with it.

Other property which a peasant wife should bring to her marriage consists of household articles: a plank for carrying loaves on her head, and a board for kneading bread and pasta, a chest for the linen, and a chest of drawers. Her mother should supply two cauldrons, one large and one small, for cooking pasta.

A man is also expected to bring furniture and other goods: the bed frame, ten chairs (two upright with arms, four kitchen chairs, and four bedroom chairs – if there is a bedroom; if not, eight kitchen chairs). He should also bring a wardrobe, a dressing table, two bedside tables, a wash-stand, a kitchen cupboard with display cabinet, a kitchen table and a coat-stand. If the house has a dining room he is responsible for the furnishings. His mother should supply cutlery, a stove and a cauldron. This list, too, is a medium length one. Many people probably do not succeed in completing it; but while others may exceed it, they are more likely to buy better quality goods. Other household articles usually come as gifts from close kin and friends, and therefore belong to the category of reciprocated gifts.

Both men and women may receive land when they marry. Not all families own land, and not all landowning families divide land when their children marry (though most do). There are no statistics to show how many people receive land when they marry, but at Caporotondo about one-third of the present owners acquired part of their land at marriage before their parents died (see below, Chapter 7).

The spouses also receive gifts at their wedding. These may be made to the bride only or to both the spouses. The gifts to the bride are usually personal ornaments; those to both spouses together are either household articles, useful or decorative, or cash. The donors of articles are usually close kin, but they may be particular friends who are especially indebted. For example, during the second world war a woman wrote letters for a neighbour to the latter's husband; when she married, the neighbour gave the couple a household ornament.

Close kin include the siblings of the spouses' parents, and their own elder siblings and any younger siblings who may be earning. Cousins may send either goods or cash, as may co-parents and godparents. More distant kin, and friends, usually give cash. Acquaintances, especially those who do not attend the wedding reception, usually send a telegram of greeting.

Gifts of cash (*bustine* – envelopes) are normally given at the reception, or on the day after the wedding. The minimum sum acceptable is 1,000 lire, and kin often give more. Gifts of objects are made some days before the reception; telegrams should be delivered during it. The following is a list of greetings, cash and goods received at the wedding of two Pisticcesi in 1965.

1. Visiting cards and telegrams

8 from friends;
1 from groom's mother's brother's wife's brother;
1 from officiating priest's father;
1 from colleague and political opponent of groom's brother;
1 from a godfather;
140 telegrams.

2. Cash

26,000 lire: groom's friends (10 persons);
25,500 lire: neighbours and family friends of groom's family (11 families);
34,000 lire: co-parents, bride's and groom's (6 persons);
30,000 lire: groom's mother's siblings (4 persons);
20,000 lire: bride's parents' siblings (2 persons);
30,000 lire: groom's mother's brother's sons (3 persons);
60,000 lire: ($100 USA) groom's mother's brother;
99,000 lire: ($165 USA) bride's brothers (2 persons)

_____

324,000 lire

3. Goods

A silver picture frame; coffee cups; a picture; a travelling clock;

ashtrays; electric carillon doorbells; silver cutlery; a musical-box and photograph album combined.

From: 3 co-parents; a respected neighbour; a cousin; a friend; a mother's brother; a mother's brother's wife's sister – all of them on the groom's side.

One pair of gold earrings; four gold bracelets; a gold necklace; a gold brooch; a gold ring; a pearl necklace. All personal gifts to the bride, from her father's sister and mother's sister; and from the groom's mother's two brothers, the mother's two brother's daughters, sister, and mother.

Apart from the sister and mother of the groom, all the people who made personal gifts to the bride also made gifts to both the spouses together. Relatively few of the bride's kin made a gift of any sort; this is explained by the fact that her family was both smaller and poorer than the groom's. Her asset, which balanced out her relative poverty, was that she had United States citizenship, and by marrying her the groom was enabled to enter the States after fewer delays and formalities than usual. No gifts were made by the girl's friends. Young girls tend to have fewer friends than young men; and this one had spent a large part of her youth in New York and was about to return there. Some gifts made by her mother's friends, 'family friends' and neighbours, are not recorded here; this list is that completed by the groom's family, recording their duty to reciprocate – a duty which extends to the bride's kin but not to her family friends.

It would be very difficult to correlate degrees of kinship exactly with the value of gifts. For example, the groom's cousins (MBS) gave an average of 10,000 lire each, while the bride's aunts gave the same amount. There is, however, as we have noted, a difference of wealth between the two families. Nevertheless, taking this into account, and omitting dollars, it does appear that there is a rough correspondence:

Groom's friends (average):	2,600
Family friends:	2,300
Co-parents:	5,500
Groom's MBs:	7,500 (+ gold)
Bride's MZ, FZ:	10,000 (+ gold)
Groom's MBS, MBD:	10,000

All these gifts are part of a series of long-term exchanges. A record is kept by the spouses (in this case by their mothers) of all gifts and telegrams received; and provided relations do not change radically, the compliment is returned when opportunity arises. Memories for these obligations are tenacious, and accurate so far as I could decently check. Gifts from close kin are calculated less than gifts from others, and are not reciprocated so strictly. For example, a return will be made when the groom's young cousins marry; but it was not thought likely that the female cousins would receive as much gold as the wives of their brothers.

The reciprocation is always made at a subsequent marriage, but sometimes this is very soon; in six months, the family above had made gifts to twenty-one different couples, most of them the groom's workmates. In some cases the 'gifts' were telegrams, but the family had already paid out 445,000 lire. If the giver of a gift is already married, it is returned when his children marry. In some cases this too may be impossible; then the return is made to a nephew or grandchild of the original donor. The obligations may thus persist for years. The gifts received by this couple, apart from those from the groom's friend's, were themselves exchanges. It is worth closing this account, therefore, with a reference to a list of the gifts which a couple married in 1940 had not yet returned in 1965. There were fourteen families on the list. Three of these were former employers, to whom it is not necessary to make a return; four were friends and kin who were married before 1940, and had no children; five were people who had moved away; one was a cousin whose husband had left her between the civil and religious marriages; and one had two young children yet to be married. The couple in question had received about sixty cards, as well as gifts in cash and kind, and had made a return for about three-quarters of the gifts they had received.

V

The wedding gifts are part of a continuing series of exchanges, made only at weddings. They should be seen, I think, as a means of meeting the immediate expenses of the wedding and of setting up a new household. The couple benefit principally from having a large amount of cash on hand when they most need it. In some cases they may not be called on to reciprocate in their turn, but this is scarcely a matter for calculation of profit or loss.

Gifts from the bride to her mother-in-law have an important

D

symbolic significance which might perhaps be called placatory: the bride is about to disrupt her husband's natal family. But since the mother-in law reciprocates with personal ornaments, the balance appears to be in the bride's favour. It is also in the bride's favour where gifts between the spouses are concerned.

Dowry and marriage settlements are more important, and present rather more problems. As a general rule the bride's property ought to be balanced by the husband's. Pisticcesi say that people ought to marry equal; but they also say that marriages are for *interesse*, and not for *amore*, and to substantiate this they will cite many cases in which 'he married her' for her property – in some cases for a particular piece of property. As I shall show in Chapter 8 many men do, in fact, rely on their wife's land to supplement their own. Women are rarely said to marry men for their wealth.

This contradiction between equality and *interesse* is only apparent, and it does not imply that men get more than they bring. Where marriage is between equals a man obviously looks for a girl who will match his property with her own; and, moreover, since men are the initiators in courtship, as well as the managers of matrimonial property, it may often appear that they benefit while women do not. But this should be seen in the context of the independence of nuclear families. In the overwhelming majority of cases the new couple becomes independent of both parental households.[8] The general practice is for parents to agree during the marriage negotiations to divest themselves of their property in quantities which are nearly equal, and as large as they can afford.

Marriages are the cardinal points of the life cycle; they are important in the domestic cycle (Chapter 3); the transmission of property determines relations between members of the nuclear family. The different ways of transmitting different sorts of property have important consequences for the constitution of neighbourhoods, and perhaps also for the relations between a man and his wife's affines (Chapter 4). In Chapter 7 I argue that bilateral transmission is a recent phenomenon in Pisticci; and I suggest, very tentatively, that together with equal marriage it may be a factor in determining what sort of system of political representation Pisticcesi have. As a major attribute of adult status, property,

[8] For exceptions, see below, Chapters 4, 9.

broadly interpreted as livelihood, is an important component of a man's honour (Chapters 3 and 4) and the amount of property a man owns has consequences for his position in dyadic ranking within the neighbourhood, and the community at large. In Chapters 8 and 9 I give quantitative information which will enable the reader to assess the relative significance of the different ways of transmitting property.

I hope it is clear that the question of the transmission of property is one which will recur all through this book. The aim of this chapter has been to put it in the context for which it occurs for each individual, that is, as an arrangement which is made in the course of courtship and during betrothal, culminating in the wedding; and as one of a set of different sorts of property exchanges. Dowry and marriage settlements are complementary transactions, negotiated to achieve equality between the spouses and to set them up as independent adults.

3

Families, Kinship and
Neighbourhood: I

I

The commonest domestic group in Pisticci is the nuclear family;
at each marriage, ideally, a new household is set up. And, as we
have seen, a large part of the transactions at marriage is intended
to provide the couple with a house, furniture and source of income.

Increasingly, families are united only in consumption, and not
in production. The members of peasant families may all work
together, but this is considered degrading for various reasons. Of
course, landless or insufficiently landed families have always had
to work for a wage, usually on a casual day-to-day basis, and
there was never any idea that these spouses and children should
necessarily work together. Now, however, it is an almost universal
ambition that children should have different – better – occupations
than their fathers.[1] Even if there should be enough land to provide
work and food for the whole family, the life of a peasant is seen
today to be less rewarding than that of a building labourer, or a
worker in the petrochemical factory. Families now tend to deploy
their members in different sectors of the economy whenever
possible.

Although 'the family' plays an important part in the teaching
of the Church and is, for example, frequently mentioned in
sermons, priests have very little to do with families as groups; nor
do families go to church together. The priests have very definite
ideas what a family should be like: a group united by love and
understanding and by the authority of the father and the obedience
of the wife and children. This unit should prevent sinful activities

[1] Compare an official elementary school text-book: 'There is much social
change because parents fall prey to stupid ambitions for their children. The shoemaker wants his son to become an accountant; the sausage vendor wants his son
to become a physician. Just imagine such foolishness.' Cited by La Palombara
(1964), p. 69.

on behalf of the Church. All of the priests saw this ideal family as threatened by the impending industrialization of the area, and tried to deflect the danger by making sure, so far as they could, that only good children got jobs in the new factories, by preaching in favour of paternal authority and, more specifically, by warning their parishioners against taking part in social activities organized by the employers. While I was there the newspapers carried reports of Bank Holiday disorder at English seaside resorts. I was consulted anxiously by priests who saw it as a portent of the end of 'the family' in their own community.

The parish priests also instituted small changes in the basis of participation in ritual. The practice of holding separate masses for the different categories of family members was replaced by one in which all should together attend a family mass. In 1965 few families did so, except on high feast days, when people would make an effort to go together; but even then men usually sat apart from the women and children, and would arrive and leave separately and at different times. Although 'the family' is given religious value, and family relationships are sanctioned by religion, families are not ritual groups.

Households are domestic groups, the members of which eat and sleep together, produce food or an income for common consumption, have a common reputation and, as often as not, are united by a concentration of affection, love and respect. A family may also be a property holding unit; and – against strangers – the property owned by any individual member is regarded as family property.

Nuclear family households are the commonest sort of domestic groups, but there are certain exceptions.

First, a few such groups are unlikely ever to become nuclear family households. A sibling who remains unmarried may be invited to live with his brother or sister and spouse. His contribution to the household economy is likely to be quite important, if the sibling is a brother, and is in any case useful. Priests do not now set up independent households, but usually live with a married sibling. War pensioners are welcome additions to a household. Such a group may be slightly richer than it would otherwise be.

Secondly, there are groups which have been or will become nuclear family households. Where there is only one person, he or she is usually the surviving member of an earlier group which has broken up through marriage and death. There are also a few people who, as unmarried children, lived with and supported their parents until they died and have not then chosen to live with another family. Such people may be of either sex, but young women do not live alone. Some widows and widowers prefer to live by themselves rather than move in with their children. All these are acceptable arrangements, made by choice. People may also be forced to live alone because they are both unmarriageable and unacceptable to their kin; they are few and, by Pisticcese standards, extremely disreputable. It is extremely rare for an individual, once established as a member of a household, to move out and set up on his own account.

Expanded households are usually those in which one parent of either spouse lives with the couple and their children. They may be formed in one of two ways: the youngest daughter may bring her husband to the parental home, or, if all the children are married and one parent dies, the relict may move in with any child. Most commonly mothers move in with daughters, for although fathers do live with their married daughters, it is rare for either parent to live in a daughter-in-law's house, and Pisticcesi tell horrendous tales about what happens when they do. Daughters usually get their houses from their parents, so that even a mother who has no legal right in the house feels less of an intruder there than she would be with a daughter-in-law; this is perhaps why daughters and not sons take the parent in. Widows out-number widowers by three to one in Pisticci, and this may be why mothers are taken in more commonly than fathers. It is my impression that widowers are likely to set up house with another woman if they can, or to marry one; but in any case they marry rather later in life and die rather earlier than women. In expanded parental households, as in the expanded sibling households I have mentioned, the additional member will co-operate in housework, and may bring in an income – a wage, a pension, or perhaps some garden produce.

There are, finally, households of a married couple without children: these are usually households in which the children have married out or have yet to be born.

There is no correlation of these sorts of household with occupation or wealth. But households which include more than one sexually active married couple do tend to be richer than the average. Pisticcesi think they should have more than one bedroom and accordingly one finds such households tend to be either in the country, where it is easy and cheap to add another room, or in a large house in town. The richest landowner in the town lived in a *palazzo* with his wife and married son, and the latter's wife and children, and two married servants. There were other households of this sort.

Nuclear family households are sometimes divided between two houses, one in town, and one in the country. They are sometimes peasant families with land, or they may be Agrarian Reform families; the same arrangement is common when the husband is a permanent labourer on a capitalist farm. In these cases the husband lives in the country and his wife and children in town; the girls may be marriageable, or the sons at school, or the wife may not like living in the country. The division may be seasonal, or for short irregular periods. There is no absolute dividing line between families where the husband goes out each day and returns in the evening and those where the husband may *pernottare* in the country once or twice a week. But some husbands sleep in the country for months at a time, and their wives occasionally bring them fresh clothes, clean the place up, and fetch food and fuel back to town.

In the country, where houses fairly often have more than one bedroom, some households have more than one married child living with the parents. Such households are productive units, in contrast to many in town; and even though each nuclear family may eat and entertain separately, there is constant to-ing and fro-ing between them, and cooking is often done in common. There were two such households at Caporotondo while I was there, and there were others elsewhere in the territory of Pisticci. There were also four sets of brothers at Caporotondo who had lived together at some time in the past with their fathers in an 'extended' household of this sort.

In each of these cases the father had not distributed his land among his sons. Normally, of course, it is divided *seriatim* as each son marries out, and part of the justification for 'extended' households of this sort is that they keep the brothers together in harmony until property can be divided at one go, when the father dies. Daughters marry out, however, and get a town house with some cash if their father can resist pressure on him to disburse land. Daughters-in-law may bring both land and a town house to the family; and this suggests that their parents have given way to pressure from their future husbands' family to forgo the usual endowment from the husband's side. If the reader cares to look now at the account of the Capece family (see below, Chapter 8, § II), he will observe that their marriages have been rather more dynastically complex than is usual. The closeness of the spouses, and of their land, may be one reason why the girl's parents are willing to do without the prestige which they earn from having set their daughters up in an economically independent household.

The daughters-in-law bring their own endowment in land to the marriage. Since they also bring a town house, the extended family is one only in the country; in the slack periods and at festivals, each nuclear family lives in a separate household in the town.

Certain arrangements for pooling labour and resources are standard. They were followed in all the extended families I knew or knew of; and were approved by other Pisticcesi. Everyone works the land; if the land is not enough to employ all the labour (as in one of the current Caporotondo cases), the surplus man works elsewhere for a wage. The wage earner puts his money in the common pool to be divided among the members. 'We work for him; he works for us.'

Consumption is sometimes common, sometimes not. The wives usually do the cooking together, and the nuclear families then either eat together or divide the food and carry it to separate rooms. Wine and tomato purée are made by the household and stored in common, each woman drawing what she needs as she needs it. Grain is usually kept in a common store, but when it is sold the proceeds are divided between the constituent nuclear families after some has been set aside to meet common needs. The nuclear families thus derive separate incomes from the common product, and may use them to buy their own land; they may also inherit in their own right; and the sons acquire land with their wives. Such land may be worked in common, and the produce pooled; or more commonly, I think, it is worked independently in the evenings.

Such extended family households are characterized chiefly by collaboration in production and by the method of dividing the product. They seem to require a forceful father and a certain level of patrimonial wealth – although in one of the disbanded households, the father appears to have rented a large farm while the scattered patrimony was more or less neglected or cultivated by other kinsmen. The purpose seems to be to delay the division of the estate, and to keep a large family work force together.

II

Any member of a nuclear family stands in two of three possible relationships to the other members: husband and wife, parent and child, and sibling. Each of these relationships is covered by legislation in the civil and penal codes. Most of the laws are permissive: a wife may require a security from her husband for her dowry if she chooses; a spouse or child or sibling may invoke the law in support of rights which it does not spontaneously enforce.[2]

The bond between husband and wife is permanent. 'Unlike other peoples', as one Pisticcese put it, 'we have no ex-wives, only widows.' Judicial separation is unheard of; and although stories were told of estranged couples, usually at least one of the spouses had emigrated. Some people live together without being married in law or in religion: I heard of seven couples, and there were probably as many more I did not hear of. One reason for these unions is that there is a legal impediment, usually a previous marriage, which prevents the parties marrying. A woman who takes a married man has usually been notoriously unchaste; a man who takes a married woman is usually a widower or poor. In one case I know of, a widower set up house without marrying because his sons wished to exclude the woman from the inheritance. Apart from the low status which attaches to these unions, an illicit couple is treated as a legitimate one; and in most cases their low prestige would probably have been the lot of each of them as individuals. It was quite common not to discover for months that a couple are not married. The matter is not discussed; there is no special behaviour towards them. One woman, whose common law husband was in prison, was visited by the police who said he had complained that she had not sent him food or money: they told her off for being undutiful and, according to her, said she should not think she could neglect him simply because she was not

[2] See Appendix II.

married to him. She then produced a sheaf of post office receipts showing that she had sent both money and food. The sergeant agreed that she was in fact dutiful, and that her lover had wrongfully maligned her. Couples such as this are probably responsible for the five or six illegitimate births each year in Pisticci.

The typical metropolitan unmarried mother is a girl who has been unchaste in adolescence or early adulthood, and who later marries. The Basilicata pattern is rather different. In Italy as a whole about 60 per cent of illegitimate births are to mothers under thirty years old, 40 per cent to mothers under twenty-five. In Basilicata only 20 per cent of illegitimate births are to mothers under twenty-five, and nearly 60 per cent are to mothers over thirty. In Basilicata, too, it appears that the number of illegitimate children who are second, third or fourth births is proportionately higher than the national figure. This suggests that illegitimate unions in Basilicata are generally those of mature people who for one reason or another find it impossible to get married; and that the unions tend to be permanent.

A husband has exclusive rights to his wife's sexual services; a wife has exclusive rights to her husband's economic services. Until recently a husband might sue his wife at law for adultery if she had any sexual relations outside marriage; she might sue him if he supported a concubine – that is, in practical terms, if his extra-marital relations constituted a continuing expense.[3] A wife is expected to bear children and rear them; she should live where her husband decides, and run their household. These are their legal rights and duties.

In Pisticci men and women have very different spheres of action. Although husbands and wives have a common interest at least in the home and in their children, there is no ideal that spouses should act together and collaborate in other matters; and in fact they generally do not. A husband, for example, does not take his drinking or political friends home, nor is it customary to call on a friend or partner without being invited. If the business is urgent the proper etiquette is to telephone (if the partner is an important enough man to have one) or to send a boy with a message. Several men told me I was the first man not a kinsman to cross their

---

[3] The law which allowed for the imprisonment of a guilty spouse has been abolished since my fieldwork; so, too, has the distinction between (women's) adultery and (men's) concubinage.

thresholds. Similarly, it is proper for a wife's neighbourhood friends to make as if to go away when her husband returns – but he may tell them to stay.

Spouses rarely go out together; the monthly market, visits to kin, weddings and the August festival are occasions when it is normal. Sometimes they act for each other. A wife may go to the Municipal Building to conduct minor business with the bureaucracy, which can only be done during working hours; a husband may buy meat or cheese on his way home. The wives of poor men may work the land. But there is little interchange of activities. This applies particularly to young and active couples. When a wife is past childbearing she is quite likely to take a more masculine role. She can go to the country without exciting comment, or take on major business with the local administration. Some formidable women use their reputation as respectable and powerful women to obtain loans.

Although the fields of action of spouses overlap very little, they are still interdependent. A wife has the right and duty to prevent her husband from damaging their common interest, and to protect him if he is in danger. A man who spends money foolishly is publicly reprimanded by his wife, who thus demonstrates to the onlookers that she is a good wife, and earns their approval. When there were fights between married men it was amazing to me how quickly wives appeared, broke through the spectators and, shouting and abusing their husbands, attempted to end the fight. It is their duty to preserve their husbands intact.

In their homes husbands are treated with respect and deference: wives do not make coffee or put wood on the fire until they are told to do so. Some women eat only when their husbands have finished their meal. But in the few families I knew intimately this formality was sometimes discarded, and the silent obedience of the wife gave way to a more balanced relation; the atmosphere might then become gay and hilarious. I should say, too, that while men often pointed to these outward forms of submission as indications of the authority they had over women and the respect they had gained from them, the women themselves would say that they found it more convenient to eat after their husbands, when they could relax over the meal. There was no question of submission they said. One woman told me she was so unused to strangers visiting that she did not know what to do till her husband told her.

It would be a mistake to assume that all spouses behave to each other in an impersonal way, meeting only in the matrimonial bed. This is a possible relationship, however, since shared activities are so few; and some men like to suggest that their only conjugal relationship is respect, not affection or common interest. It was certainly true that many men seemed more at ease in the company of other men, and women with other women, than when they were with their husband or wife. And it was my impression that in many households spouses talk mostly about family matters, weighty matters; it is only in the company of their drinking companions, or the neighbourhood women, that they talk idly and gossip.

Many Pisticcesi say that a relationship in which affection and understanding play no part is not a 'beautiful' one, however successful it may be as a marriage; conjugal love is an ideal for many, and one man told me that he and his wife were such lovers that they would wipe each others' noses. 'Marriage', a priest said in a wedding sermon, 'not only legitimates love, but commands it'. And he went on to explain that this meant the sharing of inner hopes and thoughts and fears. Another man said that the world was full of thieves and treacherous friends and injustice; sons rebelled against their fathers, and brothers quarrelled amongst themselves. The person of whom he was most sure was his wife – though God knew, he was quick to add, how sure he was.

So while it is true that a marriage can 'succeed' on a minimum of health and wealth and virtue – without, that is, the added benefits of love and understanding – most Pisticcesi expect something more, and will indeed interpret and judge their whole lives in terms of their married life: the children they have raised, the new households they have set them up in, the honour which they have maintained intact. These are necessarily co-operative tasks.

III

Parents are expected to love their children. Children are expected to love and respect their parents. Love means that they should never turn away, even when offended or hurt or extremely inconvenienced; people should make sacrifices for those they love. Respect means that children should be obedient and submissive. It is not often formulated in precise rules; and when it is, these

vary from person to person. We might say that a child's respect
is for the status of its parents, and that it should not be modified
by any particular circumstance. I was told, for example, 'You may
reason with your father, and say "Papa you're wrong"; but you
should never try to dominate him because he has made a mistake.'
This was said to me by a son whose father had been making what
appeared to be several serious mistakes. Like good and bad, or
honourable and dishonourable, the words respectful and dis-
respectful are used to justify behaviour: to gain praise for good
actions, or to justify a particular response which is not normal.
Provided the audience is sympathetic, an eloquent man can make
most behaviour seem either one or the other. Words like 'respect'
turn out to have no hard core of generally accepted meaning, but
are used for forensic justification in particular situations.

Respect, therefore, has more or less content according to the
type and intensity of the relations between people. One man, the
head of an 'extended' family, has made a will in which his two
sons are treated differentially. The son due to receive less, he
said, had stolen grain and oil from a common store; this was dis-
respectful, and although he loved his sons equally, he felt bound
to penalize the thief. At the other extreme, where sons are still
subordinate and relations are not of a very complex sort, respect
may be defined in a purely formal way: 'You should not smoke
in the presence of your father.' A widower who had recently
remarried told me that when his first wife died, some eighteen
months before, he had sent the three children of the marriage to
three different orphanages. Since then he had seen two of them
for a day each, and one not at all. They would remain in the
homes even though he could now bring them to him. He said
it was better so. How? Would they not forget him? 'No one
forgets his father: one respects one's father because he is father.'

Parents have authority over their children. This is expressed
and explained by saying that the children are the debtors of their
parents – for life, food, warmth and love. Parents will tell their
children to continue at school, or to leave school, to follow this
or that trade. One man sent his son to a seminary to become a
priest; I asked him what the son thought about this, and was told –
after a long pause – that he had never asked, but had discussed it
with the parish priest, 'who understands these things better than
we do'. Parents expect to be asked for their consent to their

children's marriages, and may recommend a particular marriage so strongly that the child obeys. Vincenzo Capece, with a young son, Carmine, remarried a widow called Pasqua who had property and a daughter, Carmela. Vincenzo had the management of Pasqua's property while she lived, but it would pass to Carmela when she died, and thence out of the family when Carmela married – unless she married Carmine. Therefore, from the age of ten, Carmela was always referred to in the home as 'the young bride' (*la novella*) or as 'Carmine's future wife'. 'We always instilled the idea', said Vincenzo.[4] When children are small they are treated with great permissiveness and indulgence by both parents. As they grow older they are expected to take an increasing part in the household work, sharing in the activities of the parent of the same sex. In peasant families boys begin work at the age of nine or so, watching animals, helping to pick tobacco, harvesting olives, and so on. Girls are entrusted with the rudimentary care of their juniors from about the same age, and are expected to help progressively with the cooking and heavy housework. These expectations conflict with compulsory schooling and with parental ambitions for their children's advancement in life. Generally speaking, the higher the economic level of the household, the later the children begin to work; and in many households the children work only in holidays and at busy seasons.

Parents sometimes regard large families as a means of social mobility. All the family should work extremely hard, and the youngest child may be pushed through law school or a seminary. This is a traditional way of achieving upward mobility, and to some extent it accounts for the relatively high ratio of professional persons to others in the towns of the south[5]–relative, that is, to the amount of work for them to do – and for the preponderance of southerners in the central bureaucracies in Rome.[6] This is not a new phenomenon, but it is much more widespread than it used to be. The expansion of state education, the creation of state agencies such as the Agrarian Reform Board, the extension of

---

[4] See below Chapter 8, §II .   [5] La Palombara (1964), p. 375.

[6] In 1959 two-thirds of all law students in Italy were enrolled in southern universities (22,622/33,169), La Palombara (1964), p. 137. Innocent XI (Pope, 1676–89) asked the Marchese di Carpio to supply him with 30,000 pigs at Brindisi. The Marchese replied he could not; but should his Holiness ever need 30,000 lawyers, he could oblige. Gramsci (1964), p. 137, where there are also more recent figures.

public works, the organization of political parties and labour syndicates and the growth of welfare agencies have all created new opportunities. Moreover, the expansion of the local economy, and in particular of the building trade, has caused many sons to follow occupations different from those of their fathers. All these influences combine to make the educational role of father less important, and take the children out of the family labour pool for long periods. The labour of children used to be regarded as one of the ways in which they discharged their debt to their parents. When the son's mobility is simply occupational, a father's duty becomes that of finding him work through his friends and allies. When the son is upwardly mobile, however, his parents become to some extent dependent on him. For mothers, this makes little difference; but fathers sometimes behave with deference, sons with superiority, towards each other. Yet the economic power of the professional son of a peasant does not belong to him alone but to his family, and it is sometimes explicitly stated that he is repaying them for their sacrifice.

IV

There are no formal rules of precedence or respect between siblings, and this is reflected in their amicable and tender behaviour towards one another in childhood and adolescence, irrespective of sex. There are no terms, in Pisticcese or Italian, for elder or younger sibling; and *fratt'mi* (brother mine), which is used as an expostulatory term of address, is not confined to siblings, but is used to anyone with whom the speaker feels at ease.

Brothers share in the general responsibility of the men of the family for seeing that sisters behave themselves and are respected. On one occasion the brother of an adulterous wife took no action when he learned that she was dishonouring her husband. When the husband eventually found out he murdered the brother as well as the adulterous couple. Public opinion, though not the law, was with him: the brother was younger than his sister, but his failure to intervene after he had been informed amounted to connivance. A brother may also intervene to protect a sister from her husband; one decapitated his brother-in-law with an axe as he leaned over a grain chest, because he had beaten and abused his wife. It should be said that this was regarded as an extreme reprisal. Informants were agreed that if a girl is seduced her unmarried brothers should intervene immediately with menaces while her father should begin to negotiate to force a marriage or procure a settlement. If the negotiations should break down it would be for the unmarried brothers to kill the seducer, and possibly the girl; they are considered more expendable than the father, who has wider responsibilities. The law, however, does not regard motives of honour as an attenuating circumstance if the homicide was preceded by negotiations: the legal assumption is that crimes of honour are crimes of passion. In Pisticci, however, it seems to be the case that such homicides are even more premeditated and preceded by agonizing calculations than, say, homicides in the course of illicit pasturing.

More commonly, brothers intervene to preserve decorum in company. Some Pisticci girls are high-spirited and may talk indiscreetly about family affairs or about themselves. One of my neighbour's daughters told me about her various *ziti* (suitors). This one was a rich peasant, that one a labourer with a building firm; but the best of all had not only got a job in the factory but was handsome (*bello*) to boot. As soon as she mentioned the handsomeness of the man, her younger brother (fourteen, against her twenty-two) told her to shut up, which she did – although she slapped him soundly afterwards. Another young man was told by a workmate that his sister had been seen walking in the street with a strange man. He downed tools, rushed home, and burst into the house shouting 'Where is she? Where has she been? Who was she with?' – all a false alarm, since the man was her mother's brother from another town.[7]

---

[7] Brothers, especially young ones, take their duties very seriously, and sometimes without much perspicacity. Older men – and the sisters – find this amusing. Here is part of a song about a girl who receives her lover at night, is disturbed by her brothers, who just miss the lover, but find the bed rumpled:

GIRL:	*Ve preje frat' non pensat' a male*
	la jattaredd' l'hav arravugghiate.
BROTHER:	*Vulite jess' accis' quante donn' sit',*
	tutte na volt' a scus' la truvat'.
GIRL:	*Vulite jess' accis' quante frate sit',*
	tutte na volt' a scus' credit'.
(—	Brothers, please don't think evil: it was the
	cat that rumpled the bed.
—	You want to be killed, you women: you're finding
	excuses every time.
—	You want to be killed, you brothers: you believe
	the excuses every time.)

As siblings grow up, marry and establish separate households they grow apart. Sisters often have houses in the same neighbourhood, and in any case remain in close contact. Brothers remain in close association with their sisters and visit them informally. Visiting between brothers is often not a casual event and may require best clothes and best formal behaviour. Such visits may be rare, on one or two feast days in the year; and sometimes brothers meet only at funerals, baptisms, marriages and so on. This, it should be noted, was called 'good relations' (*buoni rapporti*) between the sets of brothers concerned.

So common are disputes between brothers, so great the expectation they would occur, that the term is used to cover all relations not overtly hostile. Although personality was often invoked to explain these quarrels, the immediate cause, in every one of the numerous examples I came across, was a dispute about property. I never heard of one about women, and of only one about politics.

Brothers usually marry after their sisters, unless they are very much older. Men marry about ten years later in life than women, and young men are expected to help provide their sisters with a dowry and may work or become labour migrants expressly for this purpose. It is usual, then, for brothers to marry in age order at intervals of three to five years, allowing time for a re-accumulation of resources between the marriages.

This order of marriage, together with the fact that property is divided and transmitted at each marriage, makes quarrels between brothers common. By the time sons come to the altar the patrimony is already partially depleted. The subsequent divisions do not happen all at one time; and changes between two marriages in the amount to be divided may make shares unequal. In those circumstances there is little chance to settle the shares by comparing them and arguing the differences. Moreover, brothers are in competition not for a welcome addition to an already existing estate of their own, as it might be if property were divided at the death of parents, but for the means to set up a household. The prestige of a man and the excellence of his wife depend indirectly at least on the property he can bring to his marriage. Each division is the result of a complex negotiation in which not only the brothers themselves, but also their future parents-in-law, are involved. Indeed, the latter may take the larger part. I attribute the prevalence of fraternal quarrels to these factors: the division occurs

E

seriatim and the estate to be divided may alter while the process goes on; the division is crucial to the whole adult future of the brothers; and it is made not as the result only of discussions within the family, but also of discussions between the family and other families, each of which has interests opposed to all the others.[8] It may help the reader to assess my account of this pattern if he has some notion of the sort of evidence on which it is based. In the early stages of an acquaintance – when an informant was provisionally trusting, and I was anxious to make it clear that I was not prying – conversations were often about families. How many children had he? How many children had his father had? What had happened to them? He would then be quite forth-coming about his sisters: they were married, their husbands did such and such. When the conversation reached this point, there would be a pause: what had happened to his brothers? – they were married. What did they do? – they lived, like everyone else. On several occasions the wife then broke in, and said that his brothers were *cattivi*, nasty – they had been greedy and defrauded her husband. The glum silence of the husband, and the acid abruptness of the wife, gave the impression that this was a topic frequently discussed, even argued; and that it was the wife's view of the matter which prevailed. On the one hand, the man knew he ought not to be on bad terms with his brothers (and was dis-posed to call the coolest infrequent formal contacts *buoni rapporti*) and he was inclined to regard the situation as at least in part the consequence of his failure in long-suffering; on the other hand, his wife had been able to demonstrate that they alone were in the wrong. I repeat that this is my impression, a conclusion I have come to after taking part in many such conversations. It is reinforced by less frequent conversations with groups of sisters, who clearly regarded the errant brothers as resisting the just claims of their family. 'Giovanni is poorer than we are, because his brothers are *cattivi*.' It is reinforced, too, by the stated reasons for which some fathers set up extended families, and by the procedures which brothers who did not want to quarrel adopted to achieve that end.

One set of five brothers, for example, with a thirty hectare farm, had not divided it, but consigned it to one of their number. They drew a share of the proceeds according to their share of the

[8] Davis (1969, b; forthcoming, b).

expenses. Those who did not work the farm either had professional jobs or worked in the family bakery. The interest of this example is that the brothers explained their good relations by saying they had made a pact not to divide the property and to allow their wives no part in the management of any of their enterprises. They expressly saw affinal relationships as bringing discord among brothers. I have described in the last chapter how some men were able to maintain extended family households in the country. In both the extended families at Caporotondo relations between brothers and wives were peaceful, though a certain resentment was expressed against the two fathers.

It is relevant here to mention again the traditional means of mobility, by which sibling groups as a whole are upwardly mobile, working together to give one of them an education to achieve professional status. Of all professions, the most suitable was considered to be the priesthood. Unlike lawyers, priests are assured of some income, and have influence however little they earn; and unlike civil servants or clerks they can have influence without moving away from Pisticci. Well into this century there were more than thirty-five established offices for priests in the town. Each had a seat in the main church, most had a chapel in the town or the countryside, if not a parish. Parish priests, and the most senior of them, the *arciprete*, had great power; and that of a subordinate priest was by no means negligible. They were free, apart from their ritual duties, to devote themselves to their families, to politics – they could become town councillors until 1933 – to teaching, farming, scholarship or women.

Like all upwardly mobile individuals, educated and pushed up at the expense and by the efforts of their siblings, priests were expected to use their influence on behalf of their families – to get jobs for their children, to ease a nephew into the priesthood, and so on. The advantage of the priesthood over other professions is that priests have no affines to bring discord among siblings, nor heirs to the family property. This was clearly stated by my informants, who also said that a good priest is one who provides for his children from his own wealth and not from his share of the family property. There were no priests living openly with women while I was in Pisticci; but priests' children, known as *muli* – mules, infertile hybrids – were sometimes pointed out to me.

There is one further point to be made about relations between siblings. This is that I think bad relations between fathers and sons are rare because most rivalry is between brothers; and when they do arise, it is from one son's sense that his brothers have been more favoured than he has – it is an outcome of sibling rivalry. I have already attributed bad relations between brothers to the way property is transmitted; and peace and harmony between father and son can perhaps be explained by the same cause. The friction which is caused in some societies by the knowledge that a son supplants his father, and by the son's desire to do this, is lessened when the replacement occurs at the son's marriage and not at the father's death. The father knows on the one hand that his son's ambitions can be satisfied without his death, and on the other, by handing on his land, he gains prestige, and can retire into decent and respected ease in town. 'I have n children and they're all set up' is a proud boast. Pisticcesi fathers do not hope to die in harness, and are esteemed when they have fulfilled their obligations as fathers and husbands and liberated themselves from work.

4

Families, Kinship and Neighbourhood: II

I

Pisticcesi usually distinguish *familiari* (members of a family) from *parient* (relatives) and *cugìn* ('cousins'). In the strict sense *familiari* are those who belong or have belonged in the past to the same household. By imitation a child will call his parents' *familiari* his own; and by extension, a sister's husband will be referred to as *familiare*. Equally common is what I have referred to elsewhere as an exhortatory sense: a man may call more distant relatives *familiare* in an effort to convince them that his relations with them should be compounded of trust and dutifulness; 'I feel like a father to you (and so will you please carry my grain in your jeep?)' was a statement I found persuasive, and was, sometimes even in such precise terms, used to people with power and significant resources to command.

Parient and *cugìn*, with which we are chiefly concerned here, are words for both kin and affines. So far as I could tell, there is a slight tendency to use *cugìn* for blood relatives, and *parient* for affines; but there is no consistent usage: not all people make the distinction, and individuals do not always make it. *Cugìn* is a term of address as well as of reference; it was used to people who might be referred to as *parient*.

Pisticcesi certainly think of their *familiari* as a bounded group: there is ambiguity because father's brothers (for example) may cease to be *familiari* in deed; but it is safe to say that *familiari* who live in a neighbourhood are a group, and perceive themselves to be one, in contrast to 'neighbours'. Other people who share the term may belong to the group if relations with them are what they ought to be. The group is not a corporation, has no fixed positions; it is, rather, a unit of social control in the neighbourhood and, mediately, in the town at large.

Parentela is even vaguer and more confusing as an institutional form. There is no limit which Pisticcesi could tell me of. Are we *all* related, then? '*E'ovvio!*' But in fact Pisticcesi do not recognize all Pisticcesi as kin; and from time to time one might witness two people settling down to discover some link between them. People who associate regularly know whether they are kin or not. Eminent Pisticcesi have a lot of kinsmen: people claim kinship with them, across affinal links, up and down the generations. Kinship and affinity are traced out from an individual; he knows who his kin are. The people at Caporotondo could trace their links outwards from themselves so as to include nearly all other people in the district. But they often did not know the links between their neighbours.[1]

People who know themselves to be *parièt* should, quite literally, recognize each other. They should salute each other in the street, use *cugìn* as a term of address: that is all the formal obligation. However, people use the relationship for economic and political purposes, and in this case they might expect to be rather more trustworthy than two people who are not related. In general terms, kinsmen have an obligation to help one another so that when one Pisticcese wishes to impose on another, for example to introduce an inquisitive anthropologist, he tries to trace a kinship link; if this can be established, the imposition is excused, and some guarantee of good faith (i.e. the anthropologist is not a Revenue spy) is given. Similarly, if a Pisticcese wishes to extract a favour from a councillor, or to bend some bureaucratic rule in his favour, he represents this, if he can, as a tiresome imposition, yes, but the imposition of a kinsman. Favours should be done for kinsmen. The process works in reverse. Disreputable people are ignored – nobody traces kinship to them; people try to unrecognize links if someone suffers a disaster. 'My kin visit me when they want to', said one deservedly disreputable woman, and meant, they did not come when they ought to.

Comparaggio is often introduced by ethnographers as a sort of pseudo-kinship. It is a relationship created between two pairs of adults, with reference usually to a child of one pair; they are godfathers in church, sponsors in a ritual which obliges them to oversee the ritual education and well-being of the child. Pisticcesi, like almost all Italians and most Latin Americans, regard the

[1] See below, Chapter 8.

relationship as one between the adults, and sometimes use it instrumentally; mayors and ministers tend to have quite lengthy lists of *compari* and *commari*. Pisticcesi invite people to become *compari* not only at the baptism of a child, but also at first communion, at confirmation, at marriage and – less often – at first nail-cutting and (for girls) ear-piercing. When the relationship is established, the couples address each other as *cumpà* (to a man) or *cummà* (to a woman). The child is supposed to call its godparents *nunn* and *nunn'* – the same words as grandfather and grandmother. But in fact the usual term of address is learned by imitation from his parents, and he too calls them *cummà*, *cumpà*. So, gradually, all the *familiari* of each party call one another *cumpà* and *cummà*. In short, the relationship spreads out from the couples who created it, and it becomes hereditary. *Compari* are particularly trusted; they can enter each other's houses, they lend money, they do favours disinterestedly, and so on. For this reason the term 'pseudo-kinship' is slightly misleading in the Pisticcese case: it is much nearer to the relationship *familiari* have than to mere kinship. Of course, the instrumental *comparaggio* with a minister is less intense than the commoner case in which a loved kinsman is brought nearer to the household than genealogy would warrant. *Comparaggio* is a way of creating a prescriptive relationship with fairly demanding obligations. In this sense, it is like marriage; and it is like marriage, too, because it is usually instituted in a church, and is a status which has ritual implications.

That aside, it is clear that kinship is reckoned to undefined limits; as *familiari* become a group when an additional principle of residence is used, so kin are recognized when Pisticcesi use an additional principle of utility. Of course, most relatively close kin are recognized; but if one were able to draw a chart of all a person's blood relations and relatives by marriage (and their blood relations and relatives by marriage) and then draw a line round those who were recognized, the shape would be irregular, with surprising stretches in one direction and shrinkages in another. There would be, I think, a core of kin which was usually recognized by all Pisticcesi – this would consist of cousins and their spouses; and, because the rules of residence and property transmission create a group of *familiari* who tend to be recruited as husbands for sisters, I would expect there to be a common bias to that side, so that the line would include more kin of wife's sisters, mother's sisters and so on; and fewer kin of brothers' wives or father's brothers. Beyond the common centre, those who are recognized are likely to have some shared interest, usually political or economic; or to have land near to each other; or are likely to be widely different in prestige and wealth.

This brings us to a further difficulty in giving an account of Pisticcese kinship: not all kin recognize one another. A poor man will trace kinship with a remote rich man, while the latter may blow cool on the relation, in order to minimize the claims which may be made on him. He may not deny the kinship, but will give it greater remoteness than the poor man does. This is not invariably so. Some kin of this sort do have uses for each other; and politicians who need votes may adopt their poorer relatives' measurement of social distance. At one stage in my fieldwork I thought I might tie down one way in which kinship is reckoned by asking people what they would do if different sorts of kinswomen were dishonoured. This occurred to me when a girl caused scandal by going to live with an ex-policeman. She was an orphan, and the only child of a fairly wealthy Pisticcese who had married a woman from another town. She therefore had no kin on her mother's side in Pisticci; but her father's brothers were well-known and successful builders and café owners. Their first move was to sever relations. Their dead brother had been mad to marry a foreigner in the first place; and his daughter should never visit them again. 'We do not know you.' This was explained to me not as a reprisal for her disgraceful behaviour ('Poor girl, she is not a real Pisticcese; what do you expect?') for in this case reprisals would have been taken also against the girl's lover; it was rather an attempt to prevent the contamination of their own wives and daughters. The father's brothers were prepared to leave the matter at prophylaxis and considered that other measures to put an end to the scandal were not their business. After a month or so, however, public opinion moved against them: if there were no *familiari* to look after her, then it was their duty to end the scandal. Eventually they clubbed together and bought her a one-way ticket to the United States. I do not know if there were consultations between them and the girl and her lover. This episode led me to ask other people what they would do in similar cases. The answers to my questions revealed a general pattern: dishonourable behaviour by a woman should be met with

1*a*. Pisticci; Domenico Giannace in the offices of the Peasants' Alliance. Note the files of bureaucratic procedures – 'pensions' and 'various'. See p. 20

1*b*. Pisticci; The clerisy. See p. 154

2a. (*above*). Pisticci; Piazza Umberto. The poster says 'Forward the Revolution: Vietnam, Cuba'. See pp. 5, 9

3*a*. Caporotondo: Ploughing a patch of arable with trees, later sown with wheat. See p. 96

2*b*. (*opposite*). Pisticci: On the site of an old house two new ones are to be built, one above the other, facing opposite directions. The owner, a peasant and builder, stands nearest the camera, his father-in-law seated behind him.

3*b*. Caporotondo: a vegetable garden. The trees are lemon, orange and pear. Capiscum and celery grow round the well. See p. 98

4a. Caporotondo: Pietro Di Marsico (right) helps build a haystack. See pp. 102-3

4b. Caporotondo: View from the north-west

menaces against her lover's person – by her unmarried brothers, if she herself were unmarried – while her father should attempt to arrange their marriage or, if this were impossible, to get compensation. If she were married, both duties fell on her husband. Do kin have no duty then?

It was clear that men whose wives were closely related to the offending woman ought to intervene. Her sisters' husbands, for example, ought to inform a woman's husband that her behaviour is causing offence. They should strengthen his resolve to take action, and support him if they could. They should do nothing which would make them liable to criminal prosecution; and if the husband should do nothing, the best they might do would be to sever relations – though this would leave them with seriously damaged reputations. Sisters' husbands are the people most closely affected by misbehaviour, since their wives are presumed to have the same defects as their erring sister.

Most men said they would intervene if their wife's sister's daughter misbehaved; some said they would intervene if their wife's female parallel cousin were involved. Very few said that they would intervene if a brother's wife or a married sister misbehaved; it would not be their business.[2] The extent to which a brother's duty to his sister changes when she marries is illustrated by the extreme case in which a brother murdered his sister's husband because he had maltreated her: if he treats her badly, it reflects badly on the provision which her natal family made for her, and on their ability to find her a good husband. There is thus a parallel in the field of reputation and sexual behaviour to the opposed interests of affines and blood kin in matters of property.

This information, got by putting leading questions, may do no more than strengthen the general impression we have that the recognition of kinship is extremely flexible in Pisticci; but there is clearly a distinction between people who have to make some positive intervention and people who have a choice between intervening positively from goodwill and, more negatively, severing relations. The obligation on men to intervene with their wife's sisters' husbands is consistent with notions about the transmission of wickedness, and with rules about the transmission of property and residence at marriage. Where the cut-off line is drawn (wife's sisters, wife's sisters' daughters, wife's mother's

---

[2] Although the outraged husband might take a different view; see above, p. 53.

sisters or sisters' daughters) is probably related to the particular configuration of property rights and co-residence in each case. But I should add that upstanding men of honour sometimes said they would merely sever relations with their wife's sisters' husbands; and the people who would pursue wickedness into the remoter reaches of affinal cousinage were often young or themselves disreputable. It is, of course, recognized that all people, kin or not, who have associated on terms of equality with others who become disreputable, should make it clear in their behaviour that they now regard themselves as superior.

So there is probably a 'common centre', extending at least to cousins and their spouses, within which each Pisticcese recognizes kin among those available to him. Some individuals within this range may be not recognized. There are usually a number of people who, being remoter than cousins, are nevertheless recognized as kin; such recognition is often associated with possibilities of mutual exploitation. In trouble-cases a man may be forced to take notice of kinship which he would prefer to ignore, but cannot because other people recognize his kinship. For the most part this liability is operative only within the group of *familiari*, but we cannot discount the possibility that in some cases the obligations of *familiari* may also extend to *parièt*; this boundary, like others, is flexible. I have also suggested that the institution of *comparaggio* is one way in which *parièt* may be transferred into *familiari*. People who have genealogical but not residential qualifications to be members of the group of *familiari* are very similar to closer *parièt*.

II

Recent accounts of kinship in northern Mediterranean societies have tended to under-emphasize its specific nature, and – because it is used manipulatively, because kinsmen are selected, often instrumentally from the range of all those who have some relation – have classed it together with friendship and clientship; even with entrepreneurship, although the association in this case is between a type of relationship and a type of activity which may make use of either kinship, friendship or clientship. There is no doubt that this is justified. A full understanding of the politics of Pisticci, and of communities like it, requires us to understand the idioms of kinship, friendship and clientship – as well as those of organizational rank and of class[3] – and these idioms have much in common. Emphasis on what they have in common, however, distracts attention from what is specific, and that may have consequences for the formal analysis of the society which are not wholly beneficial.

When is something to be called the same as something else? Unless the objects are identical the criteria will be that such and such relevant features are common to both. The relevant features in a study of political action – particularly of political action in societies like Pisticci which are undergoing development and continual inflation of levels of acceptable living – are the instrumentality and choice with which relationships are exploited. But it were a misfortune if this should blind us to crucial differences: if it led us to argue, for example, that the study of local kinship systems makes no contribution to understanding the structure of these societies (or if the failure to study kinship led us to imply that such societies had no structure).

Kinship is ascribed: a man is born with his kin. He may discover he has more kin than he thought he had; he may wish to ignore some kinsmen and to emphasize his links with others; but none of these possibilities should lead us to think that the relationships are not 'really' kinship, or are different in kind from kinship in less complex societies. It is because the ascriptive relationship prescribes behaviour of a particular sort (trust, granting favours), that a man is pleased to discover that he has links with other men which he knew nothing of; that a man tries to sever undesired connections; that a man tries to create connections with others.

In particular, this is done by marriage strategies, by creating affinal ties which in time become kinship ties. This is most clearly apparent in the country, where marriages are marked by transfers of resources, and the resources are mapped (so that they can be taxed). To examine the provenances of a family holding is to reconstruct a series of marriage decisions. There are less tangible resources which are none the less calculated: the idea that a family is up-and-coming, and that members of it are likely to achieve positions of high prestige and influence, is a good reason for creating an ascriptive relationship with it. The evidence on which these statements are based consists of explicit statements: people

[3] See e.g., Davis (1969, b).

explain that they prefer not to intermarry with the same family in the same generation because it is a 'waste' (*spreco*). And it consists of talk about marriages: some are well devised because the affines are useful people. A consequence of strategies of marriage alliance is the pattern or 'shape' of genealogies. These are, broadly speaking, of two sorts: conserving genealogies, in which intermarriage between families in different generations – cousin marriage – recovers property which passed out of the family in an earlier generation. These are typical of the magnates of Pisticci, but also of some peasants (see e.g. below, pp. 143-5). The characteristic of these genealogies is that they have a depth of four or more generations, and a rather narrow spread. The other sort are capturing genealogies, which are characterized by shallowness – by not knowing what family one's maternal grandmother came from, for example; and they are characterized by a wide 'spread', the consequence of tracing links through affines.

It is, in fact, very difficult to write of Pisticcese land tenure, neighbourhoods, politics, without emphasizing the consequences of bilateral kinship and affinity. At the level of *familiari* this has important consequences for the composition and structure of local groups. I suggest that it may have consequences also for the question of who is recognized as *parièent*. It is because kinship is ascriptive, and because there are prescribed sorts of behaviour, that Pisticci try to manipulate their political and economic systems through their kinship ties, and in some cases to convert friendship and clientship into kinship (or affinal) ties, or into a quasi-family relationship.

III

I define a neighbourhood as an ego-centred group, the members of which live in the same area. Neighbourhoods are thus overlapping groups; they are territorially based, but are not territories. Pisticcesi use the word neighbourhood (*vicinato*); and it is a group which has daily, in most cases hourly, contact.

Neighbourhoods vary in size. Some people are more sociable than others; some people live in houses on a main route where many people pass, while others live in more isolated areas. An artizan, with his workshop in a road which was the only link between one quarter (*rione*) and the rest of the town, reckoned to know everyone in the quarter. My neighbours knew all the families in the nearest hundred or so houses, and knew all the famous and infamous people in their quarter. By consulting with their neighbours they could identify anyone they saw in the ordinary course of events within a matter of minutes.

Although Pisticcesi say that one must keep oneself to oneself, this is, in practice, extremely difficult. Most houses have no windows, so doors are left open to let in light: people can see in and out easily enough. In summer a large part of family life goes on outside the front door: all gossiping, preparation of food and relaxation. Who passes by, who buys what food, who changes his or her clothes how many times a day, is all observed and commented on. When there are quarrels within the family, when a child is beaten or a woman weeps, the neighbours know immediately; indeed, the more attentive neighbours are able to tell when such events are about to happen. Most quarrels between neighbouring families are held in the open, with shrieks and abuse which, perhaps intentionally, attract a large crowd of onlookers. One morning I was awakened at six by the noise of unintelligible but clearly angry abuse: a neighbour was conducting a quarrel with her daughter's future husband's mother, two or three streets away and some hundred feet higher up the hill. Each woman stood on her own doorstep, menacing the other through the early morning mist, across six or seven rows of sleepy houses. People should mind their own business, they say; but there is a premium on doing some things in public: justifying one's own behaviour, for example; or publicising good actions.

In Pisticci, houses are transmitted at marriage. They are given to daughters. It is unknown for a son to receive a house while his sister goes without. Parents try to acquire a house for each daughter: they may build a new one; they may add a storey to their own; they may buy one. They will always try to get houses near their own, and if two houses are available a determining factor between them is closeness to the parental home. The parental house itself will probably go to the youngest daughter if she is the last to marry; it will certainly go to her if her parents die before she marries. Occasionally, parents may move out of their house and pay rent for another one; or they may pay their son-in-law's rent. Houses are never built for sons: if the parents are so rich, they can expect a daughter-in-law who will have one of her own. Very poor people will leave their children to shift for themselves.

Ideally, then, houses pass to daughters; at least one of these houses in each family will be the house the girl's mother brought to her marriage; others will be the joint property of husband and wife, bought during their marriage. It is, however, correct to speak of matrilineal inheritance of houses in Pisticci; but while property rights pass from mother to daughter, control passes, not from mother's brother to sister's son, but from father-in-law to son-in-law. This way of transmitting houses affects the composition of neighbourhoods. There is a concentration of women related through women. It is by no means absolute or complete: many people live near to each other who are not related in this way, and many who are so related do not live near to each other. Nevertheless, it is common to find sisters living next door to each other, or only a few doors away. In the next generation these houses will be inhabited by their daughters. I did not make an exhaustive survey of housing, but I knew some twenty women who lived within shouting distance of their sisters, and knew of another ten to fifteen who lived near their mother's sister's daughters or their sister's daughters.

Some evidence to confirm this comes from a series of forty-six marriage contracts which I annotated in 1963. Thirty of these contracts conveyed houses to daughters and in ten the mother was the sole donor. Ten of the houses were contiguous to houses owned either by the beneficiary's sister (one case), by her mother (seven cases) or by her mother's sister (two cases). Unfortunately the documents record only contiguous houses, and since surnames are taken in the male line, it is not possible, using names alone, to trace any relationship between mother's sister and sister's daughter. But such evidence as there is confirms my casual notes on the subject.

These kin, and their husbands and children, are people who have more contact with each other than with anyone else. They are at once the core of the recognized kin, the *familiari*, and the core of the neighbourhood. Others of the neighbourhood are sharply distinguished as 'neighbours' (*vicini*), people with whom there is frequent but not necessarily intimate or friendly contact. The question 'which of your kin are neighbours?' can cause great confusion in Pisticci: it is like asking 'which of your dogs are elephants?' In contrast to kin, neighbours are said to observe and comment on (*triticare* = *criticare* = criticize) each other; 'You can't buy celery or change your clothes without them criticizing you' (*senza che ti triticano*).

The neighbourhood is a ranked group. People who agree that they are equals co-operate in many ways: the women lend utensils, food, even cash to each other without reserve. They may make bread together, hire a seamstress together, fetch water together, or, if one has got a cistern and the other not, give water freely. They enter each other's house, and are not offered hospitality as superiors are. Children wander in freely, and may be given food if there is any prepared. When the husband comes home the women do not go away. The *familiari* of the neighbourhood consider themselves to be equals, and are so regarded by others. The men have married into the household, and have married 'equal' – in theory, each daughter has had an equal endowment from her parents, and this has been matched by her husband's endowment from his parents. The assumption that the group of *familiari* are equals requires them to engage in mutual support and help whenever lack of resources or the hint of improper behaviour or ill-health threatens to break the common reputation and standing of the group. If the honour of one constituent family is altered, the neighbours change their treatment of the whole group; a group of families with one black sheep finds itself treated less deferentially by its inferiors, more disdainfully by its superiors. They know, too, that their reputation will be passed on, by their neighbours, to others who may consider them for partnerships in marriage, work, political manoeuvres. So the maintenance of a common standard of behaviour is forced upon them by their own claim to equality and their neighbours' assumption that they are equals. The facts that the families are largely recruited by marriages to women of the neighbourhood; that impropriety is said to be a character trait acquired through women; and that poverty and impropriety are associated, additionally ensure that the *familiari* are the people who issue warnings that some sorts of behaviour might be interpreted as dishonourable; and are the people whom one turns to in the first instance for loans and other aid.

Between unequals the formalities increase. An inferior does not enter the house without asking permission; in some cases the inferior does not ever enter the house, and in summer will stand outside the door while others sit on chairs. Loans of useful things are less frequent, and the shared activities are fewer.

It is easy to summarize the institutional mechanics of ranking and equality in a way which makes the system appear more concrete and obvious than it really is. I suppose that if an investigator were to do a series of sociometric tests, he could arrive at a description of the sort I put forward, in much less time than it took me by months of observation and questioning in my own neighbour-hood, and by cross-checking for other neighbourhoods. But, in fact, it took me some months to realize that people did treat each other differently in regular ways; that some who offered chairs for their neighbours to sit on were not offered chairs in return; that such people would also coolly say 'no' when I asked them if they were, or might become, *compari*, and so on. Quarrels which I noted down as caused by fighting between children were asso-ciated with a failure to reciprocate the term *cumpà* or *cummà*: in an account of ranking, this would be expressed as 'the x's now consider themselves to be superior to the y's'. Marriages between children whose parents do not consider themselves equal cause bitter quarrels between the parents, because one set of parents will be assimilated by their neighbours to a lower position than they expect to occupy. So it should be borne in mind that, while the abstract account of ranking presents the processes blandly and concretely, it is always a matter of slight alteration in formal behaviour, barely noticeable to the uninformed eye, but keenly felt by the participants. A man who loses or gains prestige finds that people behave differently towards him: particular people associate with him more – or less – than formerly; people who formerly did not offer him a chair begin to do so, and he may refuse a chair from people from whom he formerly accepted one.

Honour is relative, not in the sense that any one neighbourhood or other group of unequals has a limited store of honour which must be allotted sparingly, but in the sense that a man who gains prestige gains ascendancy over his former near-equals. A man who loses prestige becomes inferior to people who can now treat him with disdain. Most Pisticcesi talk about good actions with a vehement scepticism which is paralleled by the glee with which they talk about others' misbehaviour.

The practical consequence of a system of ranking based on

honour is that people who might otherwise be 'a mass of homo-
logous magnitudes, much as potatoes in a sack form a sack of
potatoes'[4] are clearly differentiated, and differentiated in terms of
their domestic behaviour. The admired qualities in a man are
diligence, calmness, lack of aggression. A man should take insults
calmly unless he is threatened as a husband or father. The good
man in Pisticci is the good husband – the man who, whatever else
he may do, avoids situations in which his capacity to provide for
his family is likely to be damaged.

Honour is a way of dividing and classifying people who are not
very different in other respects. It is based on performance of
family roles. And it is the basis of association, of forming alliances
for all purposes – mutual aid, politics, marriage, farming – from
the day-to-day to the long-term.

Honour, as a form of social control, is conservative. However
much people may be concerned to improve their reputation, and
thereby acquire new opportunities, the pressure on them is to
conform to what is appropriate in their situation in life – to their
position in their neighbourhood. Their associates comment on
and resent behaviour which appears to assert a claim to higher
prestige; their superiors resist their claims to associate with them,
partly on the ground that their rank may be lowered if they
associate too much with people whose equality with them is
uncertain. The general tendency, therefore, is to enforce conformity with particular expectations of how unique individuals and
their equals ought to behave. Reputation is formed in the neighbourhood. The assessment of honour requires a minute and day-
to-day observation of domestic behaviour: there is little hidden.
A man who wishes to impress his audience with his honour does
not recount his deeds. If a man says that his wife is under his
thumb, or that she has 'everything she wants', he is regarded with
polite indifference: if he wants to impress people he says 'Go to
my neighbours; they will tell you what sort of man I am.'
'Nothing ever touched my honour: ask the people who have lived
with me for thirty years.' 'I hold my forehead high; everyone
hereabouts will tell you so.' Honour is achieved in the neighbourhood.

The neighbourhood is a community of women: women bring
their husbands to live there; women have their close kin there;

4 Marx (1962), i, 334.

F

daughters will continue to live there when parents are dead. Women live in the neighbourhood for most of the day. It is primarily on their relations with their women that men's honour is assessed; and it is assessed by women. It is the women who do the gossiping, who observe and comment, *triticano*. The men are not there; and they rely for their information on their wives.

5

The Distribution of Land in
Pisticci, 1814–1960

Land is the main resource of the Pisticcesi; most of them derive
some of their income from it, and many derive most of their
income from it. Land has more than purely economic uses. It is
still an important component of marriage settlements, and it is
an element of prestige; it can give independence of employers and
it is a security for a man attempting upward social mobility.
Transactions in land, particularly transactions in the exceedingly
small, economically useless plots of land, are transactions in
prestige and social position.

An account of the distribution of land in Pisticci is, therefore,
of considerable interest.[1] Quite apart from the general economic
and political questions which are partly answered by knowing
how many people own the means of production and in what
quantities, it carries the description of Pisticci's social structure a
step forward. For when we describe rights of ownership, or of
use, or of tenancy, we are talking about relationships between
people. Rights imply duties and liabilities, and these must attach
to people. A hectare cannot be sued at law, nor is a boundary
dispute a quarrel with a boundary.

I

In 1814 the category of smallest properties is the largest: 615
properties under ten hectares have an average size of not much
more than one and a half hectares. The other extreme, the largest
properties, is occupied by the three men who between them have
from a quarter to a third of all the farmland; there are thirteen

---

[1] The main source of information about the distribution in 1814 is the *Catasto*
(Cadaster) of 1814. This is in the state archives at Bari. Appendix IV gives details
of how the Cadaster is organized.

properties of more than 100 hectares, and these constitute 45 per cent of all the farmland, 83 per cent of all private property.

About half the territory of Pisticci was public land. For the most part this was communal demesne (*demanio comunale*). Pisticcesi had rights to gather firewood and wild fruits, and on some of it, but probably only a small part, they had rights of pasturage. The rest, from which they were excluded, was rented

TABLE 3: Pisticci: Properties by size, 1814[a]

Size of property ha.	Properties no.	%	Amount of land ha.	%	Private property	Total property %
A. Private property						
0–10	615	91	934·38	7·3		3·9
10–100	48	7	1,177·72	9·2		4·9
100–250	4		655·78	5·1		2·7
250–500	3		1,048·94	8·2		4·4
500–1000	3	2	1,781·13	14·0	83·5	7·3
1000+	3		7,197·07	56·2		29·8
Total	676	100	12,795·02	100·0		53·0
B. Public property	—	—	11,348·79	—		47·0
Grand total			24,143·81			100·0

[a] The farmland of 1814 was measured as about 25,134 hectares (62,835 *tomoli*). Allowing for loss of land by adjustment of boundaries, this is a surprisingly accurate measurement. The tables here are based on a total of approximately 24,143 hectares (60,259 *tomoli*). The reason for this difference is that the microfilm which I had made in the State Archives at Bari turned out to be defective and to have gaps and double exposures.

out by the administration as winter pasture for the herds from the mountainous interior. The income from winter pasturage formed the greater part of the commune's revenue even into the twentieth century. Other possible sources of revenue, such as a local land tax, were never fully exploited in towns where the administration was in the hands of the local landowners.[2]

The next table, also reconstructed from the 1814 Cadaster, shows who the landowners were.

Some remarks about how this table was made are necessary. The

[2] Voechting (1955), pp. 110–17.

Church is distinguished from the priests in the Cadaster. Lands which I have lumped together under the heading 'Church' are attributed to churches (e.g. SS Pietro e Paolo) or to chapters (e.g. *Mensa arcivescovile di Taranto*). Priests' lands are attributed to individuals (e.g. *Dom.* Pietro Mastrangelo, *Sacerdote*). Nobility is distinguished by a title. The category 'Local Magnates: Professional' include all those landowners who are given professional

TABLE 4: Pisticci: Properties by categories of owner, 1814

Sorts of owner	Persons		Amount of land	
	no.	%	ha.	%
State and commune	—	—	7,289·60	30·2
Church	—	—	4,059·19	16·8
Nobility	5	0·8	8,387·96	34·7
Local magnates: Professional	12	2·0	965·59	4·0
Local magnates: Priests	26	4·4	333·84	1·4
Artizans and overseers	46	7·7	117·44	0·5
Other men	427	71·6	1,823·59	7·6
Women: Spinsters and wives	17	2·9	104·64	0·4
Women: Widows	63	10·6	135·72	0·6
Heirs of 80 dead people	?	?	926·24	3·8
Total	596	100·0	24,143·81	100·0

titles (e.g. *Notaio, Medico, Dott. Leg.*). In some cases the clerk has not always recorded the professional title; but where the name was unusual and appeared sometimes with and sometimes without the title, I have counted all the land as belonging to a single person. Where there was doubt, the land was counted under the heading 'Other men'. The same criteria were applied to the category 'Artizans and overseers'. In consequence, the category 'Other men' probably includes land which belonged to people who were educated and rich, even if they had no formal professional title. Consequently twenty people with large holdings find themselves classified with 'Other men' for lack of a more precise identification, and their inclusion in that category inflates the figure of the amount of land. If all these twenty large holdings were classified under other headings, the figure of 1,823 hectares would be reduced to about 700 hectares (3 per cent of the farmland) and the number of 'Other men' to 407. These men are the peasant landowners of 1814. So far as women are concerned, I have distinguished people clearly identified as widows (*vedova*) because it is quite likely that many of these held land not in their own right but as relicts of their deceased husbands. The only women who clearly did hold land in their own right are the spinsters and wives; there are but seventeen of them. The Cadaster makes no distinction of rank, profession, trade or sex between dead people, so that I am unable to redistribute the land among other categories, as I cannot say how many heirs there were. Finally, there are more properties in Table 4 than there are individuals in Table 5 because some individuals belong to more than one holding group, and have more than one legal personality. Wherever possible I have reduced legal persons to individuals.

II

By 1946 most of the public land had passed into private hands. There were 2,939 hectares of public land in the Cadaster, of which

TABLE 5: Pisticci: Properties by size, 1946

Size of property ha.	Properties no.	%	Amount of land ha.	%
0–0·50	1,528	37	170	1
0·50–2	1,508	36	1,736	8
2–5	623	15	1,979	9
5–10	187	5	1,281	5
10–100	180	5	5,073	23
100–200	17		2,431	11
200–500	9	1	2,787	13
500–1,000	4		3,084	14
1,000+	2		3,675	16
Total	4,058	99	22,216	100

Source: INEA (1947) Tav. I, p. 18.

2,637 belonged to the commune, 63 to the Church, 225 to the state, and 12 to a charitable institution.[3] Approximately 2,500 hectares of the commune's land was distributed in one way or another within the following two years.

The number of smallholdings has increased from about 400 to

[3] INEA (1947), Tav. VI, p. 47.

more than 4,000 although the average size of these is no bigger – 1·35 hectares – than it was in 1814. At the same time the amount of land held in properties of more than 100 hectares has only slightly declined (the apparent increase is due to the inclusion in this table of the public lands specified above). In 1814 there were 676 properties and 596 people with rights to them: in 1946 there were 4,058 properties, and 8,981 people with rights to them.[4] The pattern of minifundia – in which most people own some land, large numbers of people share rights to smaller numbers of property, and a significant proportion of the properties are very small – is often taken as typical of the immutable peasant culture of the south: in Pisticci on the contrary, it is no more than 150 years old.

The most recent development in the distribution of property in Pisticci is the Agrarian Reform of 1950. The Reform Board acquired some 2,270 hectares by expropriation and purchase, and by 1961 had distributed this to 308 families. There were 251 farms, with an average size of about 7 hectares, and 57 supplementary holdings with an average size of 5 hectares. The land redistributed was about 9 per cent of the farmland of Pisticci, and it came mainly from the larger holdings: these diminished, too, as landowners divided their estates among their children in order to reduce their liability to expropriation: normally, the management of the estates remained unified.

III

The changed distribution of property could be accounted for, on the figures alone, by saying that the big estates were subdivided through inheritance while the smallholdings were swollen with communal demesne. During the nineteenth century there was a general movement to convert public into private property, and the laws required that it should go to the landless. But what happened seems to have been rather more complex.

Certainly the big estates were divided among heirs. The sub-division was accelerated after 1880, however, when the nobility were forced by economic pressures to sell off their land in bits and pieces to a number of local families which rose to dominate the town in this period. The sons of muleteers and overseers, brothers of moneylenders and priests, the heads of these families became

[4] INEA (1947), Tav. VIII, p. 60.

first of all the tenants of absentee noble landowners, and then gradually bought their land from them. By 1935 these men had become the largest landowners in Pisticci and the most prosperous citizens; allied by marriage to similar families of neighbouring towns they together gradually acquired ownership and control of most of the land in the Metaponto plain. They were efficient and innovatory farmers who are reputed to have shown more consideration for their employees than the noblemen they ousted. Thus, the decline in the big estates cannot be accounted for simply in terms of inheritance; there was a replacement of one set of proprietors by another.

The smallholdings, too, were not made only out of the demesne. Part of the land came from the medium estates, belonging to the magnates of 1814 and broken up after 1880 when the professionals were gradually ousted from local economic dominance and political power by the families which took over the estates of the plain. The properties in the ten to one hundred hectare bracket were largely created by poor men who bought up land wherever they could. In this case, too, an account of the changes in terms of inheritance and the distribution of demesne would conceal important changes in the composition of the class of magnates in Pisticci, and important processes of social mobility among the poor.

The demesne land was not distributed only to the landless, although a large part of it was: the magnates usurped some and bought some.

Up to about 1870 there are continual statements in the Cadaster that land was restored to the commune by order of the Intendant of Finance. This is generally usurped land, but how much was and how much was not restored is hard to say. The loss of communal land by encroachment was an ancient problem. The commune, for example, acquired land from the Archbishop of Taranto in the 1790s, and found it had also acquired a seventeenth-century lawsuit with the Marchese Ferrante. In 1818 the lawsuit (over an alleged usurpation of 650 hectares) was revived, and the case ended, seventy years later, with the restoration of sixty-five hectares by the Marchese's grandchildren.

In 1865 Giovanni Minnaia, a belligerent member of the town council, moved that a commission of inquiry should be established to deal with usurpations, and claimed to have evidence that seventy-two landowners, whose land bordered on the demesnes, had encroached upon them. Eleven of the names he gave were those of councillors; three were priests; and the corporation of Pisticci clergy and the Archdiocese of Taranto were also named. An inquiry was made – not by a commission but by the Intendant – and some land was restored to the commune. The commune's records of this are fragmentary. It was, perhaps, an embarrassing business – eleven of the twenty-five members of the council were accused – and it is also very difficult to trace these restitutions in the Cadaster. I did find eighty hectares restored a few years later, but I am not confident that these were all. In general, it seems that most usurpations were small piecemeal affairs and not large-scale robberies such as Ferrante was accused of. Nevertheless, in the course of a century the effect on the demesne was probably not negligible.

It is possible that the sale of demesne was directly connected with usurpation; there is some evidence, admittedly slight, that the authorities, rather than insist on the restoration of usurped land, would settle for a cash payment. After the unification the local administrations were permitted to auction *spezzoni* – odd bits and pieces – provided that the proceeds were earmarked for public works projects. Once again it is hard to say with precision how much land was disposed of in this way, but there is a more or less complete record in the Pisticci archive of the auctions of 450 hectares put up for sale in 1872 to pay for a road to the railway station. The *spezzoni* varied in size from half a hectare to over forty. Half the land went at first offer and at first bid for prices only one or two per cent above the reserve. The rest was either bought later by private treaty, or on later offer, the reserve being lowered in the meantime. The reserve prices were known to the bidders.

The 440 hectares which were sold were bought by twenty-seven men. Only one of these (27)[5] was illiterate. The others were both rich and educated. Since there were educational and property qualifications for holding office the buyers included many politically active men and their close associates. These included the mayor (11), an ex-mayor (14), two members of the executive junta (5, 8), two ordinary members of the council (3, 23), three close kinsmen of councillors (2, 4, 21), four ex-councillors or future councillors (22, 25, 15, 10), and two priests (12, 19). The

[5] Numbers in brackets refer to the ordinals in the accompanying Table 6.

thirteen men who were active in politics or closely associated with politicians bought rather more than half the land up for sale, and seven of them were on the list of usurpers presented to the council in 1865 by Giovanni Minnaia. In only one case does there seem to have been a real dispute over a lot: this was between two men whose lands adjoined the lot, and it appears that each may have usurped part of it and wished to acquire legal rights to the whole. The price paid at the end of the bidding (by the deputy secretary to the commune [18]) was nearly 270 per cent of the reserve. It is interesting to speculate whether other buyers too might have been acquiring legal rights to land which they had already usurped; under the threat of action they might well have been prepared to pay a quittance. The general lack of bidders is by no means inconsistent with this.

But although the sale of demesne land was clearly a matter for the rich, it should not be thought that this was illegal; nor was it illegal for a buyer to be closely associated with the council. In a small town it was almost inevitable that some of the people rich enough to buy the land should be on the council. The twenty-seven buyers, none the less, were about one-fifth of the people rich or educated enough to be on the electoral roll (there were probably not more than 170 electors in 1870) so that the buyers who were also politically active were no more than 8 per cent of the electors. It may be that 52 per cent of the *spezzoni*, the amount they bought, is an unduly high proportion.

Usurpation and purchase were the chief methods by which the magnates acquired communal demesne; they were relatively few in number, but they never became massive landowners, as did those who supplanted the nobility. Probably, at a rough guess, no more than 1,500 hectares were involved. The sales occurred only between 1870 and 1900, and usurpation seems to have become unimportant as more and more land passed into private hands. Individual owners may have been more on their guard than the councillors were, against one another's depredations. In fact the nineteenth-century administrators commonly argued that redistribution was the only way to counter usurpation. On a minor scale the usurpation of public lands continues: for instance, country roads still have a tendency to become narrow paths; but this is of no great significance.

The magnates did not acquire much land through the legal

TABLE 6: Pisticci: Sale of demesne fragments (*spezzoni*), 1872–97

Year of purchase	Name	No. of items	Amount of land ha.	Difference between 1872 reserve price, and price paid lire	%
1872	1 Angiolone Nicola	1	29.12.81	+50	+2
	2 Castelluccio Canio ⎫ 3 Viggiani Domenico ⎭	2	4.60.59	+40	+12
	4 De Franchi Alessandro	1	12.19.18	?	?★
	5 Di Guilio Giovanni ⅓ ⎫ 6 Laviola Francesco ⅔ ⎭	1	9.69.15	+20	+3
	7 D'Ursi Francesco	1	19.80.19	+20	+1
	7 D'Ursi Francesco ⎫ 8 D'Angella Leonardo ⎭	2	31.65.58	+40	+1
	8 D'Angella Leonardo ⎫ 9 Delfino Nicola ⎭	1	11.77.68	+40	+1
	9 Delfino Nicola ⎫ 10 De Franchi Giovanni ⎬ 8 D'Angella Leonardo ⎭	1	36.30.71	+50	+2
	11 Franchi Nicola	4	13.42.24	+137	+12
	12 Lazazzera Giambattista	1	4.82.51	+20	+10
	13 Quinto Vincenzo	1	8.32.60	+20	+3
	14 Rogges Giovanni	2	42.76.10	+40	+1
	15 Panetta Vitorocco	1	9.81.31	+20	+4
	15 Panetta Vitorocco ⎫ 16 Boraccia Giovanni ⎭	1	20.77.86	+20	+1
	16 Boraccia Giovanni ⎫ 17 Giannone Nicola ⎭	1	10.75.82	+20	+6
	18 Conti Raffaele	1	0.66.91	+307	+270
	19 Gaeta Pietro	1	4.54.16	?	?★
1874	5 Di Guilio Giovanni	1	9.48.87	?	?★
	20 Durante Canio	1	37.44.10	?	?★
1876	21 Barbalinardo Leonardantonio	1	30.45.24	?	?†
	22 Barbalinardo Donato	1	7.15.00	−109	−20†
	23 De Pirro Giuseppe	1	35.69.77	−470	−20†
	24 Giannantonio Guiseppe	1	33.11.58	−502	−13†
	25 Pastore Nicola	1	4.48.75	?	?†
	26 Lazazzera Francesco	1	4.74.37	−179	−55†
1897	27 Lavecchia?	1	7.15.03	+4	+7†
	Unsold	1	2.37.89		
	Unsold	1	8.48.84		
	Unsold	1	1.65.95		

★ Sold by private treaty   † Sold at later auction

redistributions. In other towns throughout the south of Italy it is clear that magnates falsely pretended to be landless labourers with a right to their quota of demesne, or intimidated men who received land and forced them to give it up, or gave them credit to pay taxes and make improvements and then foreclosed. The evidence is that these practices were not common in Pisticci; and this is in part because the magnates were able to acquire the land they wanted in other ways.[6]

IV

The first redistribution (*quotizzazione*) of demesne in small lots (*quote*) to landless labourers was in 1810-14. There was then a long interval until the 1860s when, with a new central government and a new local administration, the passage of common property into private hands became a regular occurrence.

The method of distribution was relatively simple. An area of demesne was designated as due for *quotizzazione*; it was surveyed, the plots were marked, mapped and numbered. The landless were then invited to apply, and their applications were vetted by a committee which included both local representatives and provincial officials of the central government. The applicants were placed in order on a scale of need, which was assessed on the basis of their lack of other resources and the number of their dependents. Only as many people were selected for the final lottery as there were plots. Names were drawn from one urn and plot numbers from another, so that there was no question of favouritism in the allocation. The new landowners were given a contract of emphyteusis, with limited property rights.[7]

[6] For a fictional account of these manoeuvres, see G. Verga, *Mastrodon Gesualdo*. Rossi-Doria (1958), p. 199 ff. cites the classic sources. See also Voechting (1955), p. 415 ff. Voechting accepts the view of a recent writer that in a few communes *quotizzazione* was 'successful'. Villani (1955) gives a brief review of recent research, and questions the criteria on which the controversial judgement of success is based.

[7] Emphyteusis: a classical contract generally giving a tenancy and including an improvements clause. It was one way in which new lands were brought under cultivation in ancient Rome. Technically, therefore, it is incorrect to speak of the Pisticcesi as *owners*. But in practice, although the additional tax to the commune is still called a rent (*canone*), the present-day holders in emphyteusis behave in all respects like owners. For the first twenty years of the tenure, however, there was an embargo on all transactions in the land; and if the *enfiteuta* abandoned the land, or failed to pay the taxes, the land reverted to the commune.

The fairly simple procedure was controlled from the prefect's office at Potenza, and by the central government. The document which conveys the land to the new owners is a royal decree, signed by the king; at all stages reports were sent to the prefect setting out, for example, how many children each applicant had. After the distribution the central control was maintained; the archives contain many letters forbidding, for example, that a plot which had reverted to the commune should be given to another applicant until the whole procedure of advertisement, selection and lottery had been repeated. The procedure was designed to prevent exactly what did happen in other towns: that the land should pass into the hands of those who had no legal right to receive it.

The official records are, of course, silent about such illegitimate procedures as may have been successful; but the evidence for the relative legality of the Pisticci redistributions is not only the negative evidence of the records. No big landowners today, nor descendants of the big landowners of a century ago, hold land of the commune in emphyteusis. In the area Caporotondo, which I studied in detail, and which appears to be typical, it is true that a few artizans and the son of a notary received land; but they are not more than 5 per cent of all the recipients.

Between 1814 and 1950 about 9,000 hectares were redistributed. It is difficult to be very precise, but parcels of up to 1,200 hectares were allotted in 1860, 1866, 1878, sometime during the 1880s, in 1897, 1910, sometime again during the fascist administration, and in 1946-8.[8] By 1950 the largest single piece of land farmed as a unit which belonged to the commune, was about 200 hectares. Virtually all the demesne had passed into private ownership in lots of between a half and one and a half hectares.

[8] The INEA study (1946), Tav. VII, p. 54, records 2,637 hectares of communal property. Of this between 1,100 and 1,200 hectares were distributed next year; these are in the *contrada* known as Bosco Salice. A further 1,300 hectares or so were distributed in tenancies. The reason for this was that, in the generally agitated times, the local administration was unwilling or unable to withstand pressure from the more prosperous peasant landowners for a share in the last remaining substantial demesne, while on the other hand the law forbade the administration to give land in emphyteusis to those who already had some. Tenancies were the solution. But in effect these are owned by the tenants. They inherit them, buy them, and sell them; they pay taxes on them, but they do not pay rent; the collection of rent, as one well-placed observer remarked, would require a political courage which no Pisticcese politician possessed. This land is at Feroleto.

The redistribution of demesne brought no financial advantage to the commune, but rather a loss;[9] and the commune was run by men who were very conscious of the need to conserve revenues, but who nevertheless regularly approved the redistributions and on occasion agitated for them. Some idea of the motives which lay behind this contradictory position may be gained from the following council resolution, which records negotiations with the prefect and the Forestry Commission over the allotment of the demesne at Feroleto. The resolution was adopted in 1865, but the prefect successfully opposed it and that land was not distributed until 1948. The rhetoric, the parenthetical vehemence and determination to press every point home, but obliquely, make the document obscure in places, difficult to translate, and reveal the urgency which the councillors attached to the proposal. They argue a technical point first: these lands are not forest land, which would have excluded them from *quotizzazione*. But they go on to suggest that public safety demands the appeasement of the poor; that the Forestry Officer has a vested interest in prolonging the debate; that there was a precedent – a discussion about the same land in 1812 (a very weak argument this, since the commune

[9] The redistribution had certain immediate effects on the local economy, which was, in any case, increasingly liable to inflation after 1890. First, the revenues of the administration dropped considerably. The capitation fee for winter pasture had been the commune's chief income, and this was not balanced by the special taxes on emphyteusis. The administrators, rather than vote a tax on land, retrenched on services. There was continual difficulty, for example, in finding candidates for the job of municipal midwife; and the financial history of the Municipal Band is particularly chequered between 1890 and 1915. (See a typical debate in *Delibere* 27.11.1902, when the band was described as '*Forse l'unica istituzione civile che esiste nel paese*'. See also *ibid.* 23.2.1896.) The new school building, proposed and reproposed after 1865, was not built until well into the second decade of this century. Some idea of what the schools were like meanwhile is given by a speech of Cllr. Lazazzera in 1896 (30 May): 'The rooms in the boys' school are more suited to wineshops than school. The girls' rooms are like punishment cells: they are so lacking in light and air that in winter one of the teachers . . . is forced to teach school in her own home. It is a real crime . . .' The council approved another new scheme unanimously 'including', the clerk noted, 'Cllr. Franchi, whose assenting vote was, however, given subject to the condition that no further burdens would fall on the taxpayer'.

By 1920 the finances of the commune were in terrible disorder. The obligatory expenditure imposed on the commune by the local administration laws came to more than the revenues. There was not enough money to pay salaries, yet the new administration had sacked one man and created two temporary posts for their followers in his place. The rents for the remaining demesne were too low – as the Royal Commissioner, appointed after rioting had broken out, proved: he was able to raise them by a total of 1,200 lire.

had resisted the *quotizzazione* then); that *quotizzazione* would be an effective remedy against usurpation; that it is becoming very expensive to conduct the argument; that the poor are poor, and the distribution would remedy this; and finally, that it would oil relations between the monarch and his people.

. . . The aforesaid Council:

CONSIDERING that the Lord Governor has objected that the two demesnes . . . are forest land – which in reality they are not, in terms of the Ministerial Circular of 14 Feb. 1827, since the vegetation does not produce materials either for ships nor for farm machinery, and hence does not constitute forest; and

CONSIDERING that the internal peace of the towns of the southern provinces depends precisely on the distribution of the demesne lands, and hence that *minutiae* (which are the chief substance of the reports of the Forestry Officers who, seeking only their personal advantage, harden their hearts to the distress which any delay would cause) should be overlooked; and

CONSIDERING that the Ordinance of Masci [of 1810] states the rights of our poor citizens in these demesnes; and

CONSIDERING that even if the said demesnes were forest (which they are not) it would be reasonable to do as this Council has done, namely to project the *quotizzazione* of these lands as the most efficacious way of saving this Nation from [illegible word – ? *rivers*] of blood which might flow, and, indeed, have flowed in other towns of this province: whence, rather than delay the present *quotizzazione*, the Council, representing the will of the people – in whose behalf this present declaration is made – demanded that the work should be carried out, and demanded this in order to avoid every damage which might [otherwise] be done to country property, today continually menaced, and to persons; and

CONSIDERING that the two demesnes Salice and Feroleto are in part arable and in part scrub; and

HAVING REGARD to the usurpations which have occurred to the detriment of the aforesaid demesnes – the work of the agents of the Sig. Marchese Ferrante, who has already appropriated 24 *carri*, circa; and

CONSIDERING that the substance of the commune – the principal care of this Council – is wasted by repeated surveyings of the demesnes, which brings no benefit but to the surveyor – and the Council protests that from today it will meet no requests for payment which the Forestry Officer might advance

RESOLVES

that it is a simple necessity that the two demesnes, which are already divided into plots, should be distributed: and this as much for the benefit – oft proclaimed since 1808 – of the poor, who are, unhappily, still poor, as to sweeten the sentiments between people and Sovereign: the which can only occur if the demesne is distributed.

There was perhaps some justification for the arguments about public order: in the early years of the unification bandits and Bourbon partisans wandered more or less freely throughout Basilicata, and passed through the territory of Pisticci, though without approaching the town. Nor is there any doubt about the poverty of the greater part of the population. But the remainder of the arguments, severally pettifogging and grandiose, vary from the specious to the disingenuous.

V

What were the consequences of the redistributions? Some are quite clear, others more doubtful, still others speculative. Quite clear is the fact that nearly half the farmland was converted from pasture to arable: it supported a larger number of people, at any rate in the short term. Secondly, as we have noted, property in land was much more widely distributed – and the acts of redistribution followed on at fairly regular intervals between 1860 and 1950, while the population of Pisticci doubled. The point here, I think, is that the increasing population was accommodated, at least in part, and production of food was increased as a result of the distribution of land.

On the other hand, there is no doubt that the quotas of land were very small (1·2 hectares is generally a maximum size), and capital investment was minimal. There were no funds to provide capital equipment, and water, houses, trees were only slowly established. Wooden ploughs were used exclusively in the area until the 1890s. Thirdly, therefore, although production increased it did not increase enough, on the whole, to make the new owners independent of other sources of income.

The significance of this is made clear by a comparison of the growth of population in Pisticci with that of the rest of Basilicata (see Table 1). Pisticci's population increases nearly twice as much as that of the whole population of the region (100–225 against

100–125). Since we lack figures for comparative birth-rates, we may attribute this increase in part to a possibly higher birth-rate after the redistribution of land; but mostly to the fact that there was little emigration. As property was widely distributed, a large part of the population had an economic interest in the town; and the terms of emphyteusis debarred the new owners from making transfers *inter vivos*.

It has been argued that the land policies of the nineteenth century destroyed a certain sort of social structure never precisely defined, but in which feudal lord and peasant formed a cohesive social unit, and in which poverty and oppression were tempered by the traditional obligations and liabilities of the feudal lord, as well as by the benevolence which he might feel for 'his' men. It is doubtful whether this was ever a true picture of a feudal society; but even if it were, the contrast between old lords and new magnates (*borghesi rurali*) is in many respects misleading. There is very little justification for calling south Italian society 'feudal' as late as the end of the eighteenth century[10]; and it seems that the 'bourgeois oppression' of the new landowners and magnates was not so much due to their disregard for the traditional obligations of their position, but to a general economic change.[11]

None the less, it does seem permissible to argue that relationships between peasants were altered as a result both of the redistribution of the demesne and of the growth in population: we are entitled to assume, I think, that competition for additional resources became more acute than it had formerly been. And this was competition either for additional land or for other income fragments.

We may speculate that the effect of increased competition for land, as the available common land passed finally into private ownership, was greatly to increase the importance of the market (in land and in produce) as a means of social differentiation. On the one hand, by a combination of skill and good fortune some peasants acquire land from other peasants, and the gradations of rank become more numerous. On the other hand, simply because there are more ranks, it becomes possible for individuals to pursue calculated strategies of social mobility. Where the gradations are many, the differences between them are narrow, and mobility between them is easier than when they are few. Of course, the

[10] Winspeare (1883), *passim*.   [11] Rossi-Doria (1958).

G

mere existence of a market was not a sufficient cause of the changes in social relationships I am positing; the possibility of mobility depends on the existence of positions for people to move into. Here I shall note that in the first decades of this century, as the remaining pieces of demesne were distributed, the state began to extend its control over local affairs and the number of high prestige jobs increased: there was an increasing demand for clerks, schoolteachers, doctors, policemen and, latterly, land reform officials, which was met to some extent from the ranks of the richer peasants.

TABLE 7: Pisticci: Categories of house owner, 1814

Category of owner	Owner no.	%
Men	837	77
Women (married, spinsters)	79	7
Widows	145	13
Heirs of 27 dead people	27	2
Total	1,088	99

Finally, we may note that women did not, in effect, own property in 1814, and that they do in 1963. Although dowries were given with daughters in the eighteenth century, these were mostly paropherns: only the richest families could afford more, and when they did they usually gave a sum of money.[12] Women did not own houses to any great extent. The 1814 Cadaster yields the following data: there were 2,196 taxed buildings with 1,131 owners. Forty-three of the owners were corporations (the Church, various chapels, a charitable foundation, and the commune). The remaining houses were mostly owned by men.

The present distribution of houses is not known; such evidence as I have suggests that a large proportion – say, more than three-quarters – are owned by women. Nor do we know the sex of the present owners of land throughout Pisticci. But at Caporotondo (see below, Chapters 7 and 8), more than half were women.

The conclusion to be drawn from this is that the mode of

[12] Some marriage contracts dated between 1600 and 1920, for towns near Pisticci, are published in Bronzini (1964), pp. 213–51.

transmission has changed since 1814. On the one hand, it has become bilateral: both sons and daughters get property from both their parents.[13] On the other hand, a large proportion of this is transmitted not at the death of the parents, but at the marriages of the children; and, finally, daughters now get town houses, at marriage, to the exclusion of sons. There are strong reasons for supposing that these three changes are associated; and also for supposing that they are connected with the wider distribution of property in the society. What these connections might be, it is difficult to say with conviction.[14]

The redistribution of demesne lands had some direct consequences. It was itself part of a complex of changes: increasing government intervention in the local community and increasing integration of the local community in the national society and the national market. Pisticci is still in many ways an isolated, idiosyncratic society with its own dialect, its own marriage customs, religious cults, myths and traditions. But the town has become much more heterogeneous in the last century. The growth of population has created a demand for houses, and these require builders, carpenters; the number of retail shops has increased; so have the numbers of schoolteachers, clerks, bus-drivers, street-cleaners and lawyers. These occupations have been filled in part by the children of peasants who received land from the demesnes. At the same time, the peasantry has not remained an undifferentiated mass; as we shall see in the next chapters some families become richer, others poorer, in the course of time.

The population is more differentiated by occupation than formerly: it is a safe guess that most (say 95 per cent) of the working population got their main living from agriculture in the period before 1860. In 1961 this proportion was 56 per cent. If we assume that, of the total population of 6,000 or so in 1814,

[13] The legal position in 1814 is obscure. According to Salvioli (1930, p. 538) the law enjoined bilateral inheritance. The Pisticcesi clearly did not practise this. Abignente (1881, p. 177) pertinently remarks that 'Royal power, so careful and solicitous about fiefs . . . never concerned itself with the personal rights of individual citizens, and almost had no dominion over the mass of the population.' Whatever the law on fiefs may have been (and Salvioli, p. 538 contradicts Abignente pp. 184 ff.) we have no idea what the legal position was in Pisticci in 1814. Since the customs of Naples enjoined bilateral inheritance, we may assume that Pisticcesi could have practised it if they had wanted to: the pattern of Cadaster indicates that they did not.

[14] For a lengthier statement of reasoned perplexity see Davis (forthcoming a).

about 2,500 people[15] were of working age, we know that only about 600 had land of their own. The greater part of the labour force therefore consisted of agricultural labourers, dependent on wages for a livelihood. Some of these certainly held positions of authority, as overseers; or positions of responsibility, as muleteers; some, too, would have been employed permanently, others would have been journeymen. By the middle of this century the number of different types of occupation had greatly increased; and we know that some are regarded as more desirable than others. It is only in the last years – since 1950 – that the status of peasant has come to be generally undesired, so that, for example, young peasants find it difficult to get brides. Nevertheless, peasants rank themselves in terms of wealth, and it is now possible to differentiate between jobs which do or do not give one an easy life. I would suggest tentatively, therefore, that it is reasonable to say that the redistributions of demesne lands are associated with a general introduction of 'market' criteria for ranking people. Of course, it is always difficult to argue from bare figures to social behaviour but the imprecision of the early figures should not be regarded as an added difficulty: they are underestimates.

When I speak of the introduction of market criteria I mean to contrast ranking on impersonal economic grounds with ranking in terms of moral qualities. There is no doubt that people are now assessed in both ways in Pisticci, and to a certain extent these ways permeate each other, so that honour is associated with wealth, poverty with dishonour.[16] What I suggest is that this is a relatively new phenomenon, coeval with increased government intervention, penetration by national markets, and the apparent change in rules for the transmission of property.

The significance of these changes is brought out by considering their effects on forms of political representation. In a society in which economic and social organization is increasingly integrated by the market, and in which property can be conserved through marriage and transmitted from generation to generation, we might expect to find something like a class structure, even if it were of a rudimentary sort; homogamy, bilateral inheritance,

[15] This is almost certainly an underestimate. The active population is now 45 per cent of the whole population. Greater life expectancy and longer schooling have caused this proportion to decline over the years.

[16] Davis (1969, a).

ownership and market integration are all components of the class systems we know.

But there is little concentration of property ownership; and the new owners of the means of production are generally poor men who rarely employ labour and who have great difficulty in accumulating capital. Moreover, the absence of specific economic roles has the consequence that social groups are not distinguished in terms of their skill, scarce or common as they may be; they are not differentiated in the way in which in our society actuaries, say, are from building labourers. In some sense all Pisticcesi are peasants. With very few exceptions all are trying to create and maintain successful *combinazioni* of separately inadequate, under-capitalized and peasant-like enterprises. The system of social differentiation is dyadic and ego-centred; it is a ranking rather than a class system. People are ranked not only on their wealth, but on their virtue; and people of widely different prestige and power have similar values and life-styles. It is in this context that the phenomenon of patron-client relationships can be understood. They are alliances between individuals who are substantially similar: patrons and clients work within a common framework of ideas about their respective rights and duties, and each recognizes the strength of the sanctions which can be invoked by the other party. The conflicts between patrons and clients are different in kind from those between worker and capitalist; and the association creates ties between people who would, in other societies, be divided by class.

Thus equal marriage, bilateral inheritance and market organization are not enough in themselves to create solidarity between Pisticcesi belonging to the same economic category. The inadequacy of resources, the diffuseness of economic and political roles, and the absence of any segregation of rich and poor in residence or recreation, have perhaps tended to make alliances between people on different economic levels a more effective means of securing a livelihood.

6

Work

The next three chapters are based on a study of Caporotondo, a *contrada* of Pisticci. A *contrada* is a rural area of no particular size (Caporotondo is rather more than 400 hectares) which has a name and a boundary. Census data are collected by *contrade*, but I found no other administrative use. The names are frequently used, however, and all Pisticcesi knew them; those who lived in the area knew the exact boundaries. There are fifty officially recognized *contrade*, but Pisticcesi use rather more names than that – about seventy.

Caporotondo lies to the east of the town, and some six to ten miles from it, between the edge of the escarpment overlooking the Cavone valley and the main road from Pisticci to the sea. It used to be demesne land, and was distributed in 1866. I lived at Caporotondo for nearly three months in the summer of 1965, having previously spent some time in the Municipal archives, the Provincial archives, and the Cadaster office to collect material about the past. The information on land tenure which I collected in this way is complete for the period 1930–62; and the information I got by living in the area and making visits there after I had moved back to town is sufficiently extensive for me to be sure that I missed no important patterns of activity and that those I record are given here their proper relation to one another.

Although Caporotondo is a small area – 400 hectares of the total 23,000 hectares of farmland in Pisticci – I do not think it is untypical of the other peasant farming areas. I chose to live there partly because it was a middling sort of place as regards distribution of crops, numbers of houses and shacks, and, as I suspected and later confirmed, differences of wealth between the people who own the land. There are no landowners, but there are rich and poor peasants. I was able to compare much of the contemporary information about the area with the experience of my neighbours in town, who had land in different *contrade*, just as, indeed, I was

able to check much of my information about the town by conferring with country neighbours who had houses in different *rioni*
within the town.

I

At no time did I make a systematic study of attitudes to work.
The following remarks, unsupported by any statistics, summarize
my notes on the subject, which are the result of observation and
conversation.

Peasant farmers often complain of the arduous nature of their
work; they point out that the hours are long, they work in all
weathers in the open, they work a long way from their homes,
they have to get up early and return late. They get dirty and
sweaty. My own observations and my fair-weather participation
permit me to confirm this account. I have never seen men work so
hard at physical tasks.

But the work is not regular. There are long periods in the
winter when there is little to do; and many people are unemployed
for a large part of the year. Peasant farmers with horses have to
travel out to the country every four days or so to gather fodder
for the animals. They may tidy their land while they are there,
but they do not do much else. People work extremely hard for
part of the time and then have little farmwork to do. The
effort required for any particular task is greater than it would
be if they had lived on the land they farmed, or if they had
machines.

Work is not valued for its own sake. I never heard any Pisticcese
say anything that could possibly mean 'work has an intrinsic
value'. The poorest labourer concurred in saying that work was
degrading, physically and morally. 'Ah, so you've come to see
me working: it's hard work. And do you think I'm a person?
(*cristiàn*).' Peasants resent the fact that some people do less work
than they do and live better for it, but this does not lead them to
say that being a peasant is better than being a schoolmaster; rather
the reverse, for they share the common value that it is better to
live from rents, or commerce, or a profession, than to work with
one's hands. 'We produce food. How would they do without us?
And how do they treat us?'

Pisticcesi peasants do not say they have a special calling to the
land: they do not think of themselves as in a particular communion

with nature. I often heard men contrast their position with that of a lawyer or teacher saying that however hard peasants work they depend immediately on God, but it never seemed to me that they derived satisfaction from this observation; rather, they were resentful at being singled out in this way. 'You, Englishman, have your vocation (*avvocazione*); and I have mine. But mine's condemnation (*ma la mia condannazione è*).'

Farmwork is a hard lot; infinitely preferable is to work as a clerk or professional or to do no work at all. Pisticcesi peasants, perhaps because they live in the town in close contact with *galantuomene*, have detailed knowledge of how they work and live, and they generally share the gentry's picture of themselves: yokels (*cafùn*), dirty, poor, uncouth, speaking incorrectly.

Peasants justify work in two ways. Most will say that to produce food is necessary: where would we be without it? Richer men, who are self-sufficient or nearly so, also value their work for the independence it brings: 'I produce my own food. I don't stand under anyone.' Their livelihood may be precarious, but at worst they expect to scrape by without being at the beck and call of an employer. People who do not have enough land try to acquire more, 'to be independent'.

Work is also justified in terms of the family of the man who works: 'If it were not for my family, I'd not be wearing myself out (*non mi sacrifico*).' The ability of a husband to support his wife and children is as important a component of his honour as his control of his wife's sexuality. Independence of others, in this context, thus implies both economic and sexual honour. For, in the past at any rate, the wives and daughters of farmworkers were fair sexual game for the husband's employer or his agent. So, too, were the women who work in weeding gangs on the big estates.

Women may work in the country either as labourers or as their husband's helpers. In either case this can cast doubt on their chastity. They are in danger from their employer; or their husband cannot support them by his own efforts. A man who is weak in performing his economic duties is often assumed to be feeble in exercising his sexual rights – that is, neither to satisfy his wife himself, nor to prevent other people from doing so. As I have shown elsewhere,[1] a man who has seduction in mind approaches

[1] Davis (1969, a).

a woman whose husband's prestige and power in the community is inferior to his own; and it appears to be the fact that poor women are seduced more often – though my knowledge of cases is necessarily limited – than prosperous ones. Thus, whether because people know what men's intentions are, or because they know that poor women really are seduced more frequently than others, there is a complex and close association of poverty (a sign of which is working in the country) and *disonestà*. Poor women are assumed to be loose women; and the same explanation is frequently given why women work, and why they commit adultery: 'She does it for her family.'

Women who work the land are thus involved in a potentially dishonourable activity. There are certain exceptions to this: women who are not sexually active are immune, and many do in fact work. Women are exempt from suspicion at harvest times and when they go to the country for illegal purposes, such as killing pigs without paying the appropriate taxes. Some women, mainly wives of clerks or professional workers, may go into the country 'for the air' for a couple of weeks in the summer. They do not generally work, but they may follow rustic pursuits without incurring suspicion.

Work, then, is not regarded as having any intrinsic rewards. Men work to produce food and some cash for their families. Women who work are considered immoral. In spite of this it would be wrong to think that people get no pleasure from doing particular tasks well. Men with special skills and knowledge quite clearly enjoy exercising them and displaying the results. One man was famous for a herbal remedy which he rubbed into mules with pulled muscles. Other men have commoner and more workaday skills. Pruners and grafters took great pleasure in well-pruned olive trees, and would call my attention to the finer points and explain the principles of technique. Pisticcesi distinguish about fifteen ways of grafting trees, and practise most of them. Arguments and discussions about grafting sometimes end in practical displays taken to lengths far beyond any horticultural end. Fig trees which bear a few fruit of six or seven varieties which ripen in succession, or plum trees whose branches bear alternately red, yellow and green plums, are the results of such displays, in which people appear to find considerable pleasure, freely admitting that there is no practical use.

II

Most farming at Caporotondo is extensive, and the main crops are staple foods, farmers keep back enough for their own needs and market the rest.

The main crops are grains, pulses and olives; in addition, most farmers have a small vineyard and a vegetable garden in which they produce solely for their own consumption. Some have citrus orchards, and a few produce enough pears to market what they do not eat.

TABLE 8: Caporotondo: Land use, 1962

| | Cadastral units[a] | | Amount of land | |
	no.	%	ha.	%
Arable	258	35	200.10.11	47
Arable with trees	221	30	147.22.02	34
Olive grove	87	12	49.13.77	10
Vineyard and olives	46	6	11.04.52	3
Vineyard	73	10	16.08.28	4
Other	48	7	5.03.31	2
Total	733	100	428.62.01	100

[a] i.e. *Particelle*: see Appendix III.

The figures in Table 8 are the official ones. They give only a rough guide since changes of use, which have been quite common during the last decade, may not be registered for some years after they have been made. Arable land can be used for grain, tobacco or pulses. 'Arable with trees' is a translation of *seminativo arborato*, and generally means that olives are not planted too thickly for crops to be grown between them; these are a poor man's olive grove. 'Other' here includes the small vegetable gardens, farm buildings and rather more than one hectare of land which I could not identify.

Although these figures are approximate it is clear that the land is chiefly farmed under grain or pulses and olives, and that this is extensive farming. As can be seen from the synoptic chart (Appendix VI), a hectare of grain requires less than 150 hours' work a year, a hectare of olives – and Caporotondo olive groves average 0·6 hectares – requires about 260 hours' work by men. There are some 400 farming units at Caporotondo, so that extensively farmed buildings are very small: if size were the determining factor these would be best farmed intensively.

Intensive farming requires continual attention, which is incompatible with the economic structure I outlined in Chapter 2. It also clearly requires, as a glance at the synoptic chart will show, a great deal of women's labour, which is incompatible with the social controls operating to keep women in town. Other important considerations are the lack of capital (buildings, machines, tools and equipment) and the tenurial system, described in the following chapters, which would make capital investment risky – though I have no doubt that, were grants for such investment made available, people would soon take care to define their property rights more precisely. I should also mention that land changes hands fairly rapidly. Both these points are discussed at greater length in later chapters. Water, too, is scarce. About half the farms have wells, but these get very low in the dry summers, and there is certainly not enough to irrigate everyone's land.

The state of the market for intensive crops is also a limiting factor. Peasant farmers have long-established relationships with grain merchants to whom they sell whatever grain and pulses they do not keep for seed and food; and they pay on account for bread or the services of smiths, cobblers, bakers and barbers – all of whom will accept payment in kind in Pisticci. Some grain merchants deal in *sansa*, the residue of olives after pressing; and there is generally a ready market for excess oil or for the olives themselves. Wine is produced mainly for home consumption, and any surplus is sold to other Pisticcesi who do not have enough, rather than to merchants.

An assured market exists for tobacco, which is grown under a system of licences, so that production is pre-regulated. But the market for all other crops is more or less unstable and intermittent. A man who had invested in about a hectare of vines for table grapes, for example, found it impossible to sell them. They were not, if the truth be told, very good quality grapes, since he had abused his copper-sulphate sprays; but in 1965 even the big landowners were retailing their own grapes in market places in the neighbouring provincial capitals. The Caporotondo table-grapes grower tried in vain to get a job as doorman in the petrochemical factory where, he wrongly supposed, he would be allowed to set up a stall. Excess oranges are sold to merchants who come from Potenza, Taranto or Matera, who pay low prices and do not come at all unless the crop has failed in more accessible areas. The green-grocers in Pisticci town prefer to buy in the wholesale markets at Taranto and Bari, where prices are lower, qualities are standard, and large quantities can be bought from one man. Only one Caporotondo peasant had a licence to sell in the daily retail market at Pisticci, and she said she had to 'fight' her husband to get goods to sell.

The commonest way of selling small amounts of produce is to put them in a basket on a chair outside the door of a house in town, when passers-by will bargain for them. Some producers have a permanent arrangement with a kinswoman who lives on a busy street in town; some peasant women become known as always having an egg or two for sale, or figs in season, and so on. Eggs, fruit, artichokes, tomatoes, lettuce and celery are all sold in this way.

There is in any case considerable resistance to the idea of offering small amounts of surplus produce on an impersonal market, and in some cases of selling them at all. Quite often I was told that crops were *per uso familiare e per complimentare gli amici* – for family use, and for giving away to friends. When I tried to buy wine for my own use I was sometimes told that the producer would not sell it but would give me what I wanted.

Doubtless, if Caporotondo peasants increased their production of, say, oranges, merchants would find it profitable to come there more regularly. Doubtless, too, outlets could be found for the chili and capsicum peppers which grow luxuriantly round the wells in the gardens. But the peasants think they cannot intensify production until they have water and a market. The economic risk and the social penalties are too great: if the crop falls below what are often imprecisely calculated expectations, then they have no other resources to fall back on. If they spent more time in the town and less in the country they would find other opportunities more readily.

III

Some summary details of the amount of work required to pro-duce various crops are given in the synoptic chart in Appendix VI. This was drawn up from an extremely thorough study of costs and returns in the Metapontino, where the land is flat, and the conditions, by contrast with Caporotondo, ideal.[2] A few remarks may be useful.

First, it must be made clear just how much more labour cash crops require. A hectare of cereals needs about 150 hours a year, of which five hours are women's work. A hectare of tobacco requires 2,650 hours a year (women, 1,170) – nearly eighteen times as much. A hectare of table grapes requires 1,020 hours (women, 580): of oranges, 1,100 hours (women, 200). For all those who do not live in the country a labour input of these proportions is out of the question. Travelling time between Pisticci and Caporotondo is a minimum of two hours a day. For those who have land in other areas as well, the problem is even more acute.

Secondly, those who do live in the country are either very poor, in which case their margin for survival is very narrow and they cannot take the risk of farming intensive crops; or they are past maturity, their children are married and their economic needs are not at the maximum. People at their maximum level of economic activity have sexually active wives, marriageable daughters, and sons at school: they generally live in town.

Thirdly, the intensive crops require high concentration of labour at certain periods of the year, and this is beyond what most Caporotondo peasants can supply, even if all their family worked. It would be necessary to employ labour, something which, except in the special case of tobacco, discussed below, is difficult to do. Peasants at Caporotondo employ each other for casual or specialist 'favours', not for continuous work. To employ a permanent labour force of labourers is not economic, for there are long periods of repose; moreover, the owner would have to live on the land to supervise the work.

Finally, rather more than half the landowners own land in more than one place; and most of the farmers rent, or sharecrop, or have other-use rights to land which is scattered – sometimes as much as five hours' journey away. The problem of intensively cultivating a distant plot is insuperable.

IV

Peasants mostly work alone. But even a man who is always solitary in his fields is co-ordinating his efforts with those of his

[2] De Benedictis and Bartolelli (1962).

wife, who takes care of domestic work and perhaps of chickens and the vegetable garden; and he is likely to organize his farm-work so that it fits in with other jobs which he may have. One Caporotondo peasant worked in Stuttgart and returned for the grain harvest each year, staying for the winter sowing and then going back to Stuttgart. Such long-term organization was not common. On a smaller scale men fitted in jobs as building labourers with the seasonal variation of work on their land. A water-diviner cum well-digger at Caporotondo made that activity fit the demands of his quite intensive farming.

In addition, Caporotondo peasants do farmwork for other people. Some of the very poor go out to work for others for a wage; they take their own equipment, and are paid more if they supply their own mule or horse. They are paid the standard rate, or near to it: this was up to 2,000 lire per diem in 1965, without a mule. Five or six men did this from time to time; the arrange-ments were made some time in advance, and nearer to the general period indicated the employer and the labourer would meet and settle on the day: the man then arranges his own work to fit in with the job. It might last two or three days. One man worked for three or four different employers in this way, and seemed to think this normal. The employers are rich peasants, migrants, or the socially mobile: clerks and schoolteachers who have land and who farm it themselves rather than let it out or sharecrop it. So far as I could tell there was no competition between the Caporotondo men for these jobs, and most of them were outside Caporotondo.

Others did 'favours'. A favour is usually a hard day's work using a special skill, such as pruning, in return for a nominal sum. An expert pruner welcomes the attention which a request brings; and he works only for people who are close to him: neighbours, friends, kinsmen. A man who mistakes his standing with the expert, and asks him to prune for him, is brusquely told 'I am not a labourer.' Food is expected and offered, but it is not part of a contract. The payment for a favour of this sort is up to 1,000 lire; the standard rate in the market is about 2,500 per diem. The expert works entirely at his own convenience. I knew three men who did favours.

Thirdly, there is exchange labour in a strict sense. This is restricted to certain sorts of task: tidying a vineyard, pruning, weeding. The characteristic these have in common is that they permit sociability in a way in which ploughing, say, does not. Exchange labour is restricted also to certain sorts of relationship: friends, kin who are on good terms, close neighbours. The exchange is made on a short-term basis – men work together for a day or two, spending half the time on the land of each. The return is roughly equivalent, and amounts are small; if one party should renege, the other can cut his losses without much pain. Most people who exchange labour do so as a part of a continuing relationship; but the exchanges are made for particular jobs, and I never heard it suggested that any of them would agree to work together over a period of time. How often a man has help, how often he helps others, depends on the number of kinsmen he has in the area and how sociable and friendly he is. One man, very sociable, with a reputation for honesty and gentleness, reckoned to be working for someone else for a day or half a day in a week and therefore worked in company for one or two days: he had half a dozen exchange partners.

At harvests labour may be recruited in any or all of these ways. Kinsmen are called on, women are brought from town, and there is a general exchange of labour. Grain is now almost universally harvested by machines; in 1965 only one man at Caporotondo cut his wheat by hand. The grain is mown by a simple harvester, bound and stooked by hand; it is then carted to a stack and threshed by machine. No combine harvesters are used in the hillier and broken areas of Pisticci.

The threshing machines are owned by Pisticcesi who are generally successful master-builders or garage owners. The machines and their crews travel round the *contrade*, going wherever there is most work. The peasants, therefore, try to club together; they want the threshers to come as soon as possible, and it will come quicker the larger the group is. When it arrives, all the members of the group help each other. The machines are old and need a mechanic and one or two labourers in attendance. Otherwise the work of feeding the machine from the stacks, carrying the sacks from it to the weighing machine, and making the stack of chaff is done by the group of peasants. At each stack the women tie the sacks, watch the weighing, and prepare food for the harvesters. The women do not travel from holding to holding as the men do. The thresher is paid for by a percentage of the grain threshed, usually 6 per cent.

Pulses are harvested by hand. The plants are pulled up and left to dry out in the fields. They are then carted to a threshing floor – no more than a hard patch of beaten earth – and threshed by trotting a blindfold mule or horse round and round. They are winnowed, first by tossing them in the air with wooden spades, then by centrifugal force in a sieve suspended from a tripod, then by hand, when stalks, pods and dirt are removed. I once saw a three-ton lorry driven over and over a mound of chick peas, but that was as far as mechanization advanced: they were then winnowed in the usual way. Commonly the threshing is done by men. One trots the mule, another shovels the pulses under its feet; but if there is only one man the shovelling job is done by the women. Winnowing is done by both men and women. A wife and her mother and daughter usually do the hand sorting.

Olives are harvested by spreading a cloth under the tree and shaking the branches one by one. Men shake; women pick out the olives from the twigs and leaves, and put them into sacks. One man can shake for several women. I was ill during the main part of the Caporotondo olive harvest, and was unable to observe whether the statement that labour was not bought, and that the women were all close kin to the tree-owners, was accurate.

Problems of organization arise mostly during the grain harvest. The exchange of labour for company on one-man jobs is rarely urgent or difficult to arrange; it is fairly easy to maintain reciprocity, or to cut losses in the rare cases when this fails. But at grain harvests disputes are likely to arise. The machine-owner is more likely to send his machine to a large group, and the larger the group the more likely it is to include people who are not under obligation to other members, or who are even at odds with some of them. The order in which stacks are threshed may be a matter of dispute, since people are assumed to do less work after their own grain has been threshed than before, and if in fact they do slacken their efforts, this is a cause of resentment. It is difficult to maintain strictly equivalent returns since some men have larger stacks than others. Poor men, with small stacks, are inclined to grumble if richer men appear to work less for them than they did on the rich men's stacks. Each man assumes the direction of the work while his own grain is threshed, and occasionally this circulation of command is resented. The different tasks do not require equal effort; and while people generally shift around and try to see that everyone does all the jobs, this is another matter which is likely to be raised once a quarrel starts. Pitchforks and knives are to hand, and are used to threaten, if not to fight, when tempers rise. Fights are stopped by the women, who rush to the scene and disarm their husbands, pushing them apart.

Harvest quarrels are common: the machine-owner's labourers reckon to see three or four each year, and indeed appear blasé and unconcerned when they start, welcoming the occasion for a rest. The quarrels do not subside easily, partly because they are not concerned really with the harvest, but split the group where it was least cohesive in any case. Incidents may occur for some months afterwards; in another *contrada* an aggrieved man took his horse to foul another's well. This quarrel was typical in that it involved a preliminary dispute about the composition of the group in the first place, before the thresher came, as well as past harvest quarrels, a boundary dispute, and a suspected seduction some decades previously.

Harvesting groups are the largest ones – up to ten heads of households with their women and children – that a Pisticcese peasant is ever likely to work in. They are created in response to the exigencies of the machine-owners, and include people who are not on good terms. Apart from these, some men work with others in brief exchanges of labour which are part of a continuing relationship. Some men work for others: experts do favours for people who are socially close to them, and poor men work for a full wage for people who are not socially close.

V

Pisticcesi say that tobacco is very difficult to grow. It certainly requires a lot of work, by both men and women, and some capital equipment: the wooden frames for drying the leaves, and a store-house where the pungent bundles of dried leaves are stored for a couple of months, from September to November. It seems probable that it is these considerations, rather than the inherent difficulty of growing the crop, which make Pisticcesi reluctant to cultivate tobacco.

It has been grown on a large scale in the area only since the war; and the usual method is to import families of labourers from the Lecce area, where the crop has been established longer and where

H

the inhabitants are said to be experts. The Leccesi come on a sharecropping contract (*compartecipazione*) which is barely profitable to either party, although the Leccesi value the perquisites. Landed property is not widely diffused in the Lecce area and the people are perhaps poorer than those of Pisticci.

By terms of the contract the landowner has to prepare the ground, provide the equipment, all transport, lodging and some

TABLE 9: Metapontino: Balance sheet for tobacco sharecropping
(1 hectare)

	Debit	Credit
Owner		
4 Quintals tobacco		200,000
Equipment	18,400	
Mechanized preparation	103,000	
Food	16,000	
Lodging, storage	30,000	
Transport	5,000	
Total	172,400	200,000
Leccese		
4 Quintals tobacco		200,000
Food		16,000
Lodging, storage		30,000
Transport		5,000
Manual work	265,000	
Total	265,000	251,000

Source: De Benedictis and Bartolelli, 1962.

food, which is agreed as a fixed amount of bread per hectare of land for the period. Some Pisticcesi also provide salt and oil. In recent years there has been a flow of emigrants from the Lecce area to north Italy and north Europe, so that tobacco workers have become much scarcer and are in a better contractual position. They now insist on a lump sum to be paid in addition to the half-share of the harvests. This varies from 75,000 to 100,000 lire per hectare; larger sums are paid on the coastal plain.

The Leccesi families – usually, in practice, one man with all the women and children, say five to six persons in all – do all the work

except transporting water to irrigate the newly transplanted seed-
lings.

Since then the price paid for tobacco has been raised from an
average of about 50,000 lire/quintal to about 90,000 lire. But the
cost of food and labour has also increased. The Leccesi, for their
part, earn little more than a labourer's wage, while the Pisticcese
now has to pay not only the lump sum, but also their travelling
expenses. Even allowing for the fact that at Caporotondo and in
other inland areas the preparation of the soil is unlikely to be
mechanized, the Pisticcese is lucky to break even.

At Caporotondo in 1965 there were about thirty hectares of
land, farmed by a dozen families, under tobacco. Some of the
fields were extremely small – less than a quarter of a hectare – and
it is becoming increasingly common for the owners of these small
plots to dispense with the services of Leccesi. But only one man
who had more than a hectare of tobacco did not employ Leccesi.
The quality of the tobacco produced was not noticeably different.
Some of the small plots were planted in unsuitable soil, in the
shade of trees for example, but I heard the licensing authority's
inspector-adviser reprove Leccesi and Pisticcesi alike for picking
the leaves while they were still immature, dark green and shiny,
before their points began to curl. He confirmed my opinion that
no arcane skills were needed.

The men who cultivate tobacco without Leccesi sharecroppers
bring their women to work in the country. Most of them are poor
men with little reputation to lose. But richer and independent
men do bring their women: one man made his daughter, a Lycée
student hoping to go to university, pick tobacco; she was very
annoyed. Another man was forced to bring his women and a son
when his Leccesi family did a moonlight flit because they thought
the season so dry that the crop would fail.

Sharecropping is uneconomic for tobacco, and it is difficult to
explain the Pisticcese reluctance to abandon it and instead make
full use of their own labour resources. Those expert skills which
are allegedly needed amount in practice to no more than knowledge of the time to pick the leaf, although Pisticcesi do not usually
grow crops which need transplanting or spraying. They say they
do not know how to dry leaves, or even to thread them. One
man said the Leccesi were better at stooping to pick the leaves. To
some extent the explanation must be that the work does require

continual vigilance: the leaves have to be picked during August, which is the period after the grain harvest when the town's major festivals occur. During September the leaves have to be exposed to the sun on wooden frames, and put under cover when it rains; it is difficult for a man to leave his tobacco while it is outside in the sun.

It is hard to know how much emphasis to put on the social controls which are certainly applied to prevent women going to the country. It was noticeable that the few men who did bring their women out were either poor and disreputable, or remarkably independent in other matters. Their politics tended to be both radical and anti-party; one, not at Caporotondo, publicly expressed agnostic views, in a town where communists are content to be anti-clerical. The Lycée student was placated by her father, who told her he might let her go abroad for a study trip. On the other hand, those men who did not bring their women never said that they were concerned with their honour; they just said the Leccesi knew more about tobacco.

I am much inclined to say that motives of honour are the main reasons why Pisticcesi do not adopt more profitable methods. By this I mean to indicate the social penalties which follow from a loss of honour: the difficulty of arranging suitable marriages, worsened relations with affines, and so on. But I do not invoke any principle of innate conservatism. The penalties are real, and are applied by people who make no direct profit from the tobacco. It is remarkable, incidentally, how common it appeared to be for the Leccesi women and girls to have some at least mildly sexual relations with the young men of Caporotondo. Not only were there dances at least once a week, but it was not uncommon to see unescorted girls walking along the country paths hand in hand with local men. The fact that the Leccesi women – who live only in the country – are known to be accessible to some degree, may confirm the Pisticcese view that it is unsafe to take women there.

7

Fragmentation at Caporotondo

This chapter and the next are about land tenure in Caporotondo. I begin by discussing taxable rights, and in Chapter 8 go on to discuss other rights, and the way the two sorts are combined.

I

The Forestry Officer's report of 1865, before the demesne Caporotondo was redistributed, describes it as an expanse of treeless uncultivated scrub: gorse and mastic, ideal – the officer said – for allotment to the proletariat.[1] The following year the land was reclassified as arable, divided into 331 plots (*quote*) and given in emphyteusis to 361 persons. Of the plots, 271 went each to one man; and another 40 were held each by one woman, usually described in the lists as 'widow, mother, guardian'. Twenty-one plots were held by two or more people. Of the 361 people, 339 or 340 appear to have been poor and landless; they may have been favoured at the expense of others who were poorer, but they were not magnates by any stretch of the polemical imagination. There were nine others who, although probably poor, had other occupations than agriculture: there are two men identified as goatherds, one cabinet-maker, a wine-maker, a shoemaker, a bricklayer, a salt-seller, a farm steward, and a *fuochista* – who may have been either a stoker or a pyrotechnician. One man, Vitorocco Viggiani, was certainly the son of the notary, Domenico Viggiani, and two other men were probably related to magnates. Giuseppe Castellucci was probably son and nephew to two town councillors, and Francesco Cantasano was possibly the uncle of a man who

[1] The Forestry Officer's report gives the area as 510 hectares. Part of this is no longer called Caporotondo, and I took this as a good enough reason to limit the area of study. About fourteen hectares appear to have got lost – probably through mistakes in measurement and through errors in registration, as well as through my own mistakes in copying. The area of Caporotondo for present purposes is, therefore, slightly more than 426 hectares.

became a town councillor fifteen years later and subsequently mayor.

By 1962 (the Cadaster was up to date for that year in 1965) the number of properties had increased from 331 to 412, and the number of individuals having rights to land from 361 to 703. Of these only 329 (47 per cent) were men. The number of properties which were shared by more than one owner had increased from 21 (6 per cent) to 151 (37 per cent).

TABLE 10: Caporotondo: Properties by the number of owners, 1962

No. of owners	No. of properties	No. of persons
1	261	261
2	78	156
3	27	81
4	15	60
5	9	45
6	5	30
7	5	35
8	5	40
9	5	45
10	1	10
14	1	14
Total	412	777[a]

[a] In several cases a number of persons are only one individual: hence the discrepancy between the figure of 703 individuals and 777 persons.

II

In the century since the redistribution of the Caporotondo land there has, then, been a great increase in the number of people who have property rights there; and there are more properties than there were in 1865. The reasons for this, and the way it has come about, are the theme of this chapter. It is a case-study of 'fragmentation'. But this word has different meanings for different people, and a preliminary digression, in which we apply common sense to identify the realities to which it may refer, is perhaps useful.

I want to suggest that the common view of fragmentation, which sees it as a state of affairs resulting from the application of inheritance laws, is inadequate. When fragmentation is taken to mean a state of affairs, what is generally meant is that the farmers are poor, or their holdings incapable of supporting them. This may be the result of fragmentation; it may also be the result of changes in the acceptable level of living, or of changes in the structure of the local economy. It is quite common for isolated communities, when brought into contact with industrial markets, to acquire new needs for industrial products without there being any increase in the prices of agricultural produce; if production is not, or cannot be, increased, farms become 'inadequate' and farmers poorer, even though their farms may have remained unchanged in size. Moreover, the diversification of a local economy may make it possible for farmers to become part-time farmers, so that the deficiencies of their agricultural incomes can be made up in other sectors. This may be regrettable;[2] that is a value judgement which distracts attention from interesting social processes. Part-time farming is not the consequence only of irrationality in farming, but also of 'irrationalities' in the development of the local economies.[3]

It is perhaps more useful to look for a *process* of fragmentation rather than a state of affairs: when the word refers merely to a condition it may be used to blanket interesting complexities of the sort I have just sketched. The process of fragmentation is one through which holdings become both more scattered and smaller. The Italians, in fact, use two words: division (*polverizzazione*), and scattering (*frammentazione*). I now suggest that these two processes are not necessarily associated, and that neither is the consequence solely of inheritance laws.

Holdings can become smaller without becoming more scattered. Where men exclude women from inheritance, for example, holdings may be progressively divided with each new generation, but they do not become more scattered. Thus Stirling describes a society in which land passes from a father to his sons, and the fields are a genealogical map of a patrilineal sort.[4] It is worth pointing out, too, that the strictest application of a rule of male inheritance will result in a general impoverishment of a community only if the population is increasing and the total area available is limited.

---

[2] Kato (1964), p. 57 regrets it, in Japan. German *Arbeiter-Bauern* are on the whole approved of.

[3] See Davis (forthcoming, a).

[4] (1965), pp. 44–5.

Some fathers will always have more than one son, but only if this is a general phenomenon will the average size of holdings become significantly smaller.

Holdings may become more scattered without becoming smaller. Take the hypothetical case of a bilateral society with a steady population: if each child in every generation inherits half the property of each of his parents, in time the holdings will be widely scattered, but the total amount of land held will be the same as it was in the first generation. Thus a man might inherit one-eighth of the holdings of each of his great-grandparents: his sibling and his first and second cousins will have the same amount of land as he has. All that is required is that the population should remain constant. A less hypothetical case is for a man to buy land which is not adjacent to his own: his holding has then become more scattered, but bigger. And the vendor's holding has become smaller, but has certainly not increased its scatteredness; that may even have diminished.

To summarize:

1. in societies with unilineal or bilateral inheritance systems, holdings will become smaller if the population increases;
2. in societies with unilineal inheritance systems holdings will not become scattered as a result of population increase; they will, if the inheritance system is bilateral;
3. market operations may cause holdings to become more scattered and bigger, or less scattered and smaller;
4. external economic changes can cause holdings to become too small without affecting their size.

This is, of course, a purely formal argument which does no more than draw out the consequences of distinctions not always made. I think, nevertheless, it is clear that the idea of fragmentation as a state of affairs resulting from inheritance laws is inadequate. Only specific forms of inheritance law produce fragmentation either by division or by scattering, and markets can have similar effects. Before trying to put some flesh on these bones, we might query an underlying assumption: that peasants are generally so tightly gripped by the ineluctable procedures of inheritance laws they ruin themselves by dividing their land to uneconomic levels. There is a *prima facie* case against supposing that peasants, who make their living by the land, are incapable of calculating what an adequate farm might be; it is their livelihood to know. More-over, laws, and in particular inheritance laws, are often evaded. The literature on peasant societies abounds in cases of people in communities isolated from the law-giving centres of control, who manage to circumvent laws in the pursuit of their own wealth and well-being. To put these points more positively, the doctrine that fragmentation is caused by the mindless application of inheritance laws seems to exclude the possibility that people with small and scattered farms may have chosen to make them so and that, in choosing, they were enabled or encouraged by the character of the economic environment. In particular, if our townsman's model of fragmentation insists that the 'structural defects' exist only in agriculture, or only in peasant conservatism, service and industrial sectors which permit or encourage part-time involvement escape our notice.

III

In 1866 the average size of properties at Caporotondo was 1.29.00 hectares. In 1962 it was 1.01.50 hectares, a drop of 21 per

TABLE 11: Caporotondo: Variation of properties from average size, 1866, 1962

% of average size	1866	1962
Less than 25%	0	26
Less than 50%	0	57
Less than 75%	4	83
Between 75%–125%	260	139
More than 125%	57	63
More than 150%	9	18
More than 175%	1	9
More than 200%	0	17
Total	331	412

cent. In 1866 most holdings were near average size; in 1962 the range was much greater.

In 1866 each of the new properties was a continuous area. By 1962, as a result of inheritance and purchase, some thirty-two Caporotondo holdings were scattered. Twenty-eight owners had

two separate pieces of land, and four had three. Much more significant, perhaps, was the scattering outside Caporotondo. Of the
landowners at Caporotondo in 1962, 117 held land in other
contrade as well. Many people who belong to owning groups at
Caporotondo have rights to land as members of the same group,
in other *contrade*; some people who have shared rights with one
set of people at Caporotondo have shared rights with another set,
and some people who have shared rights at Caporontondo have
sole rights elsewhere.

TABLE 12: Caporotondo and elsewhere: Scattered properties, 1962

	No. of properties	Hectares
Caporotondo	32	—
Elsewhere (same association)	117	356.80.94
Elsewhere (different association)	86	268.75.13
Total	235	625.56.07

Allowing for the overlap of the two categories of 'elsewhere'
holdings there are 172 owning groups or persons in Caporotondo
who have rights to 625 hectares of land outside. Thus the original
427 hectares at Caporotondo are augmented to a total of 1,052.
Most of these holdings are under ten hectares in area.

We can now make a crucial calculation: the 412 owners at
Caporotondo are 10 per cent of all owners of less than ten
hectares, and the total amount of land they have rights to is
21 per cent of all the land held in properties of under ten hectares.[5]

It is a mistake to assume from this that Caporotondo owners
have, so to speak, expanded outwards, establishing a hegemony
over other lands; they are not noticeably richer than the peasants
of other *contrade*. Rather, the peasants of Caporotondo have been
absorbed into the intricate jig-saw of interlocking rights to land
as a result of intermarriage, bilateral inheritance and purchase.
Whereas in 1866 the new owners at Caporotondo had one-plot
properties, unfragmented, and were to that extent a discrete set
of persons, 'the holders at Caporotondo', they are so no longer.

In many cases these far-flung rights to land are not used for
farming: they are potential rights which can be made operative

[5] INEA (1947).

by activating kinship obligation. There is, as I shall show in the next chapter, a tendency for people to concentrate their holdings to some extent, and to waive or cede rights to land which is distant from their own and take land at rent which is nearer.

IV

In the second section of this chapter I argued that a distinction should be made between fragmentation which originates from market operations and that which results from the transmission of property between generations. In earlier chapters I have also described how important a part the acquisition and dispersal of property plays in the individual life cycle and in the domestic cycle; men acquire property when they marry, increase it as they approach maturity, and then disburse it as their children marry, leaving themselves some control over a portion as a security for their old age. It is of some interest to calculate the numbers of different types of conveyance and the amount of land affected by them. Fortunately the Cadaster normally records how land is transferred.

The common forms of conveyance are seven, some of which are linked – I call these supplementary conveyances. *Successione* is the transfer of property from a dead person to his heirs. It is best translated as inheritance, but it should be remembered that Italian law reserves up to two-thirds of a man's estate as legitim, to be divided among people who have a legal right to the property (see Appendix III). Inheritance by a group of heirs is followed by a supplementary conveyance, *divisione*, which is used to divide a single property among the several new owners. I translate this as division. Some groups of owners are established in other ways than by inheritance – by purchase and by gift, for example – and each division must therefore be traced back to the next preceding conveyance. Sometimes co-heirs hold the property jointly for an indefinite period, during which one or more may give up his rights by a supplementary conveyance known as *cessione*, cession. Marriage settlements are made by *donazione* (gift) and by *contratto matrimoniale* (marriage contract). Most gifts are made by owners to their descendants – only women receive land by marriage contract; both men and women receive land by gift. Purchased land is conveyed by a deed of *compravendita*. Finally, there is *riunione di usufrutto* (return of life interest), which is another supplementary conveyance. Many of the conveyances *inter vivos* (even, on occasion, purchase) reserve a life interest to the conveyor; when this person dies the usufruct is reunited with the bare property. By law, too, a widow or widower has the right to a life interest in part of the property of his deceased spouse. The primary conveyances are thus inheritance, gift, marriage contract and purchase. The supplementary conveyances are division, cession and return of life interest.

Between 1930 and 1962 there were 811 conveyances of property at Caporotondo.

TABLE 13: Caporotondo: Types of conveyance, 1930–62

Type	No.	%
Inheritance	180	22·0
⌠ Gift	200	24·5 ⌉
⌡ Marriage contract	14	1·5 ⌋
Purchase	142	17·5
Division	124	15·5
Cession	20	2·5
Return of life interest	79	10·0
Other	45	5·5
Unknown	7	1·0
Total	811	100·0

Table 13 shows that there are more marriage settlements than inheritances, and that purchases are nearly as common as either of these two sorts of conveyance. Inheritance is a more or less automatic conveyance – few Pisticcesi make wills, and the laws are applied. By far the greatest number of conveyances are between living persons: 61 per cent, in fact, including division and cession. Excluding purchases, 44 per cent of all transactions in land are between living people who stand in some close relationship to each other.

Table 14 shows the amount of land conveyed by the various methods. The figures here refer to the last conveyance for each piece of land, and show how the present owner acquired it.

Once again the percentage of conveyances *inter vivos* is high at 67 per cent. affecting 61 per cent of the land. Excluding purchased

TABLE 14: Caporotondo: How present owners acquired their land; amount of land affected, 1962

| Type of conveyance | Conveyances | | | | Amount of land adjusted[a] | |
	no.	%	ha.	%	ha.	%
Inheritance	48	11	53.91.03	13	98.27.85	23
Gift	114	27	108.95.13	26	139.89.57	33
Marriage contract	13	3	12.90.60	3	13.82.37	3
Purchase	76	18	73.29.68	17	88.44.45	21
Division	75	18	56.69.84	13	—	—
Cession	7	1	7.38.64	2	—	—
Return of life interest	37	9	45.21.41	11	—	—
(Aboriginal holdings[b]	44	10	52.96.73	12	60.64.66	15)
Other	12	3	12.44.11	3 ⎫	24.49.89	6
Unknown	3	—	3.21.88	— ⎭		
Total	429	100	426.99.05	99	425.58.79	100

[a] By redistributing supplementary conveyances to the next preceding category, as follows: 45 to category Inheritance, 44 ha.; 35 to Gift and Marriage contract, 32 ha.; 14 to Purchase, 15 ha.; 3 to Aboriginal, 8 ha.; 1.40.26 lost through error in transcription.

[b] I.e. those in existence in 1930, when the present Cadaster was begun.

TABLE 15: Caporotondo: Year of acquisition of property rights by present owners, 1962

Year	No. of transfers to present owners
Before 1930	44
1930–34	41
1935–39	38
1940–44	40
1945–49	65
1950–54	72
1955–59	77
1960–62	45
	422
Unknown	7
Total	429

land, then, 44 per cent of the land was acquired by its present owners from close kinsmen who were alive at the time. The relative weighting of the three contrasting types of conveyance is also clear from the adjusted figures: 36 per cent of the land is acquired by gift and marriage contract, 23 per cent by inheritance and 21 per cent is purchased. This illustrates my earlier discussion of the life cycle, and of the significance of the transfer of property at marriage. The fairly high proportion of the land which is sold confirms my point that, in societies where fragmentation is

TABLE 16: Caporotondo: Number of times *particelle* changed hands, 1930–62

	Particelle no.	%
Did not change hands	55	6
Changed hands once	150	18
Changed hands twice	248	30
Changed hands three times	313	37
Changed hands four times	52	6
Changed hands five times	15	
Changed hands six times	8	3
Changed hands seven times	2	
Total	843	

allegedly due to the blind adherence to inheritance laws, we should look carefully at market transactions. Perhaps more important, on a general level, than either of these is the great quantity of land which passes between living kinsmen. Land transactions are an important component of the relationships between them, and cannot be interpreted simply in economic terms.

Two further sets of figures can be usefully extracted from the Cadaster for Caporotondo. Both tend to show how long any one piece of land is held by a particular person or group.

Most properties are relatively young; and the total number of transfers in any year is fairly constant. There is thus a fairly rapid turnover of land at Caporotondo; three-quarters of all the property rights are less than twenty years old. Similar conclusions can be drawn from rather a different calculation. The Cadaster assesses tax on a unit of land called the *particella* – one piece of land

ownership. At Caporotondo
: from a few square yards to a
rizes data which show how
ferred in the period 1930-62.
particelle change hands cannot
the only slightly misleading
it regular intervals – we may
hange hands every ten years
years. I have suggested that
it to work on the land (see
is reasonable to suggest that
ic areas over long periods of

ild way, I have argued that
y the bare figures of size of
has to be seen as a historical
ocial and economic context.
tion of the domestic cycle.
f Pisticci as a whole – the
ther less than a quarter in
er of people with rights to
The average size of holding
falls to about 80 per cent of the 1866 average size, but the range is
much greater. As far as scattering is concerned, this affects 204
properties – about half – and is associated for the most part with
holdings outside Caporotondo. People who have rights there
(427 hectares) have rights altogether to 1,052 hectares. The
average man to hectare ratio, therefore, is rather better than it was
a century ago, though holdings are more scattered.

There is no doubt that this is not a modern agriculture. Not
only is the tenure system inefficient and wasteful of time and effort,
but the land is used primarily to produce food and a surplus above
consumption needs of safe, unadventurous and not very profitable
crops – grain, pulses and olives. Such investment as there has been
is mostly directed to producing limited quantities of other food-
stuffs.

Nevertheless, fragmentation is not a dead-end process of inevitable and increasing poverty. Many peasants are part-time

farmers. Some people with rights to land are shopkeepers or clerks or teachers, and their land is available for cultivation by another person. In fact, about half the land at Caporotondo is cultivated by people who do not own it. The ways in which the available land is redistributed to meet the fluctuating needs of cultivation is the subject of the next chapter.

8

Rights and Resources

In the previous chapter I discussed, under the title of fragmentation, the ways in which taxable rights to land are transmitted from one person to another. Some transmission is between generations and, in theory, given the increase in population, this should lead to progressive impoverishment. But I argue that we should assume that peasants treat their rules of inheritance as pragmatically as, say, their rules about dishonour, and are as practical about plot size as they are about growing wheat or caring for themselves. I attribute poverty not to ignorance of what the right size of a farm should be, but to the existence of a market. On the one hand this is associated with changes in the local definition of an acceptable standard of living; and on the other, some people are offered the opportunity to become rich, and are able to take it. Where resources are limited, and investment to increase productivity is minimal, this necessarily means that other people become poorer. But in any case the intrusion of the market has not been the only way in which Pisticci is more closely integrated into the national society. There has been an expansion of opportunity generally so that rich peasants move off the land into other occupations, while the poor have access to the land which they do not use, and can also work in other sectors of the local economy.

People at Caporotondo have taxable rights to more than a thousand hectares of land scattered throughout the countryside of Pisticci. They also have untaxable rights of cultivation. As tenants and sharecroppers they have use-rights enforceable at law; as kinsmen they have quasi-property rights which are informally ceded to them and which, for one reason or another, are not registered in the Cadaster. All these rights can be transformed into property rights by usucapion after thirty years' uninterrupted use of the land, and this is sometimes done. Arrangements of both sorts affect about 35 per cent of the land at Caporotondo, and are,

I

therefore, an important part of the strategies by which the cultivators make ends meet.

Pisticci is a relatively complex society, and Pisticcesi can do most
things in one of several accepted ways. What they do is to some
extent determined by factors which I am content to regard as
accidental: by personality, by having no sons, by having no
daughters, and so on. Some initial choices also set a pattern.
Matteo Laviola (below, section I), once he had 'chosen' to work
the land, had very little scope for further choice, since he was
so poor. Vincenzo Capece (below, section I), on the other hand,
was rich and could ignore the disapproval of his neighbours.

Pisticcesi say that life is a struggle. There was no one at
Caporotondo who could afford to allow himself to be carried
along passive on the ebb and flow of his domestic cycle. Not only
were they all concerned in the tactics of agriculture – deciding
what to grow, when to harvest, how best to prune a particular
tree; they were all concerned, too, to implement a strategy, to get
access to resources, and to do so in a socially acceptable way. As
we shall see (below, section V), 14 per cent of the land was owned
by women and cultivated by their husbands: whom to marry can
be a strategic decision.

This chapter, then, begins with four examples of the way
people have acquired the land they use and how they use it. I then
discuss how land becomes available for use by 'arrangement';
then present the data on the amount of land actually available;
and I end with a discussion of kinship and affinity between
cultivators at Caporotondo.

I

A PSEUDO-EXTENDED PATERNAL HOUSEHOLD

Vincenzo Capece is the head of a pseudo-extended family (see
Figure 1). Two of his three sons by his first marriage live with him
in the country. His eldest son is a clerk in the Cadaster office in
Pisticci; the second, Antonio, married his own paternal grandfather's great-niece; the third, Carmine, married his father's
second wife's daughter by her first marriage. Both these sons have
land through their wives; Vincenzo, who is seventy years old,
owns and controls the family land. Vincenzo and Carmine live
together with their wives and Carmine's children, and they eat

together. Antonio lives about twenty yards away in a house owned by his father. Vincenzo's land there is half a plot, the other half of which is owned by Antonio's wife, who received it as a gift from her family when she married.

Antonio is more detached from the extended family household than his brother, but they all work together, and he has no separate supply of grain or oil; the grain is threshed and stored, and each takes what he needs. Vincenzo's sons, about forty and thirty-five years old, were very conscious of their father's control

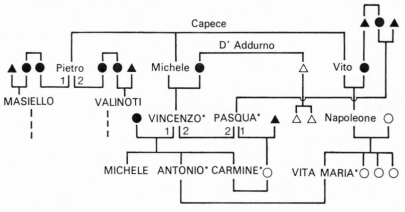

* Members of Vincenzo's household

FIGURE I. The Capece family

and while they never rebelled against it, they would explain in public that the existence of this eccentric household depended entirely on his will.

Vincenzo Capece was given 2.57.86 hectares by his father. Most of this is arable, but he had a house on a narrow strip of vine and olive grove. From his mother he got another strip of arable with olives, a total endowment of 2.87.51 hectares. He married a girl with land close to his own: the extremely valuable orange grove, house and vineyard with a well where he now lives, and some arable in a *contrada* called Tinchi. The couple thus had 3.80.69 hectares. This wife died, and the land passed to her children; Vincenzo got a life interest in the whole. When their only daughter died her quarter of her mother's land was divided between him and his sons; they own eleven-twelfths, he owns one-twelfth.

These are the only property rights he has in common with his sons. In 1934 he bought arable in two other *contrade*, at Feroleto and at Tinchi. And in 1950 he bought the plot adjoining his wife's land; he has only half this land, the other part belongs to a friend with whom he purchased it. In 1964 he bought more arable at Tinchi, having rented it for eighteen years before. The bought land amounts to 3.73.05 hectares.

He also rents three plots. One of these was, in 1965, a short-term tenancy of type becoming increasingly common: the tenant takes a fallow field for six months, pays no rent or a nominal one, and leaves the land *pulit* (ploughed and harrowed in time for winter sowing). The Capece family also grew tobacco on some of Vincenzo's land, and all this land under tobacco was share-cropped by Leccesi. Another rented plot belongs to three brothers related to Vincenzo's second wife Pasqua in a way I failed to record. This is a long-term tenancy; the brothers are emigrants to the U.S.A. The third rented plot is on a short lease, renewable yearly, from Vincenzo's mother's brother's son, D'Addurno, who is a labour migrant to Germany.

The Capece enterprise was further consolidated by three marriages. Vincenzo's first marriage, as we have seen, brought him land near his own most valuable field. His second marriage, to Pasqua, brought him more land and her daughter also had land at Caporotondo which she had inherited from her father, Pasqua's first husband; Pasqua has a life interest in this land. Pasqua's daughter and Vincenzo's youngest son were brought up to believe they would marry, and eventually did, thus preventing the loss of the Caporotondo land which would have followed the girl's marriage to an outsider. The second son, Antonio, also made a good marriage, to his paternal grandfather's brother's son's daughter, re-uniting a plot which had been divided two genera-tions before and which – through deficiency of sons – would have passed out of the Capece name. Antonio lives in the house on his wife's land and regards the re-united plots as his own; his father, however, has not removed the boundary stone between them.

To resume, then: Vincenzo Capece was endowed with 2.87.51 hectares; he got 1.82.93 hectares with his first wife; and another 1.10.53 with his second wife. His sons' marriages brought in 1.65.57 hectares, Antonio's wife's endowment being worth more than its market value because it re-united plots which had been divided before. Marriage strategies over a period of thirty years thus brought in 4.59.03 hectares. He has bought 3.75.05 hectares, and rented (in 1965) a further 3.19.54 hectares – a total of 14.39.13.

Vincenzo's 'patriarchal spirit', as his neighbours and sons called it, provides us with a good starting point for this brief series of examples. Since he insisted on running an extended family household he had not dispersed his land, and we can see the effects of a long process of accumulation unaffected by marriage endowments. His total resources, including rented land, are five times his own marriage endowment from his parents.

Cousin marriages such as Antonio's consolidate property as well as 'keeping it with the name'. Some Pisticcesi emphasized the value of this, although what the value consisted of they were unable to say. It was not only I who pressed them. Pisticcesi, too, argued about this, many were scornful – as they were in general of Vincenzo's 'patriarchal' practices. Probably the chief value of cousin marriages is that, as in this case, they re-unite land which was divided by grandparents or great-grandparents.

Although people poured scorn on Vincenzo's feeling that land ought to pass in the male line, they agreed with him that what his father's eldest brother has done with his land was shocking. This man had given part of his land to his first wife, so that, if he died first, she would have property rights and not a mere life interest. But the wife died first, without issue, and her share of the land was inherited by her sister, married to a Masiello. The widower Pietro then re-married, having now no more than a third of his original holding. 'Since he was by now quite accustomed to giving things away', Vincenzo commented sourly, he made his remaining land over to his second wife. She survived him, and when she died again without children, the property went to her sister's daughter, a Valinoti. The history was well-known in Caporotondo, where it was told as an example of the lengths a man will go to spite his brothers – giving land to wives who bore him no children. The land is now a Valinoti-Masiello enclave within the Capece territory: Vincenzo continually remembered the loss of it, and would not speak to the present-day owners.

THE DI MARSICO SIBLINGS

In 1965 Pietro, Vincenzo, Giuseppe and Maria, the children of Giambattista and Antonia Di Marsico (see Figure 2), ranged in age from sixty-five to about forty-five. As they married, each moved out of the parental home and took a part of the patrimony. This amounted to a total of 8.17.21. When he married, Pietro took 1.66.69 hectares of good land, with a house on it. Vincenzo took 0.91.55 hectares of good land and about two and a half hectares of dry wheat land, not very good, and far from Caporotondo; Vincenzo's wife's land is at Caporotondo, and has a house on it. Giuseppe, who is not a farmer, took 0.30.61 hectares of good land. Maria took 0.74.76 hectares of good land. The parents

FIGURE 2. The Di Marsico siblings

retained 2.03.60 hectares, almost a quarter of their own land. Some of this belonged to Antonia. When Giambattista died his wife retained a life interest in his land. He left no will, but expressed the wish that it should be divided between Giuseppe and Maria. The behaviour thought proper in such circumstances is that the other heirs should gracefully cede their rights to achieve the ends desired by the deceased ex-owner. What happened in this case, and it is what Pisticcesi thought quite normal, was that Vincenzo objected: he thought he had been hard done by in the settlement on his marriage, and that consequently the land should be divided between the three of them; he therefore refused to sign the document waiving his rights under the law of intestate inheritance. Antonia held on to her usufruct, and persuaded her other children to refuse to carry out a division in terms of the law. Meanwhile, she got Giuseppe to farm all the land and tried to placate Vincenzo by offering to divide her own land between him and Pietro. Vincenzo objected that the half-share which he was offered was not worth as much as the part of his father's land to which he was entitled. Antonia then divided her land between her two granddaughters, children of Pietro and Vincenzo, who were named after her. Vincenzo quarrelled bitterly with his mother, and when she died he did not attend her funeral; the other children took this as a refusal to be reconciled at any cost. Pietro and Vincenzo are still on bad terms; their father's land remains undivided; and this plot is therefore registered in the names of Giuseppe (three-quarters) and Vincenzo and Pietro (one-eighth each). Another plot, which was the common property of the parents, is registered in the names of Antonia Sisto for one-half, of Giuseppe for three-eighths, and of Maria for one-eighth. Antonia Sisto, the mother, who died about 1950, has a usufruct on all the land she does not own. There has been, in short, a moratorium on all transfers because nobody will sign the documents; Giuseppe is still farming the land.

The state of affairs in 1965 was that Pietro and Vincenzo were living permanently at Caporotondo, but avoiding each other. Pietro's town house has been declared unfit for habitation because of landslides; Vincenzo's wife's town house has been passed on to their daughter. Giuseppe appears very little at Caporotondo, and Maria lives with her husband in another *contrada*. Most of what follows is about Pietro and Vincenzo. Perhaps it should be said that both are respected and friendly men.

Pietro is voluble and outspoken; he has strong political opinions which he attributes to the communist party; his son-in-law Domenico is secretary of the communist Peasants' Alliance and an ex-mayor of Pisticci. He has partly withdrawn from active life since his second daughter's marriage: his needs are not very great, and he sells nothing but grain, which covers most of his expenses. Both he and his wife have pensions. Vincenzo, unlike his brother, speaks Italian; he is richer and more enterprising: he grows two and a half hectares of tobacco with Leccesi sharecroppers, markets some vegetables I think, and was reputed to lend money at interest.

Pietro received two plots from his father; he bought another from his father's brother with the proceeds of a three-year tenancy of a thirty hectare farm. This is all the land he has ever held – a total of 3.21.95 hectares. One of his daughters married a Genoese, and received a settlement in cash. The other, Antonia, who had land of her own from her grandmother, was given one of Pietro's father's plots, and his father's brother's plot. Pietro farms all this land, and also the land belonging to his son-in-law. He rents land at Feroleto from the commune; it is three hours away by cart. He thus now owns 0.74.79 hectares, and farms 5.78.26 hectares, excluding the unknown area at Feroleto.

Vincenzo got 3.41.55 hectares from his parents. For many years he lived on land which he rented from his wife's mother's sister; this became his wife's property by inheritance when her aunt died. He sharecrops two plots which belong to a schoolteacher, the son of a peasant, and his wife; this arrangement dates from before the war. When his children married he gave his Caporo-tondo plot to his daughter, and his dry arable to his son. He was left with no land of his own, and took various plots on short-term tenancies from emigrant landowners; these amounted to 3.52.66 hectares in 1965. Together with his wife's land and his long-term sharecropping land he thus had 5.74.49 hectares in 1965.

Pietro and Vincenzo are fairly typical of families in which property is handed on at the marriage of children. Vincenzo has no property at all now. He is criticized for taking on tenancies at his age: since his children are married he ought to retire. Since his children are married and he has no home in town, he argues, he has nothing to lose by working. Pietro, too, has handed on most of his property, and works in close association with his son-in-law: he does all the main work on all the land, and his son-in-law takes what he wants from their common store of wine, grain and vegetables.

MATTEO LAVIOLA

So far the examples have been of families which are rich by Pisticcese standards. Matteo Laviola is a poor man without enough land even to employ him. He is married, with a fourteen-year-old daughter, who lives with her mother's sister in town and in 1965 was training to be a seamstress, and a nine-year-old son. Apart from the daughter, they live permanently in the country. Matteo and his brother, Michele, worked as *una società* for ten years or so after their marriages, until 1960. They farmed their own land, rented and sharecropped other land and both worked as farm labourers. When eventually Michele broke away, he turned to building and established a reputation for himself as a well-digger and water-diviner. Matteo kept the tenancies, and continued to work as a labourer. He grows mostly grain and pulses, largely because his labour resources are so small. In 1965, for example, he grew 0.30.00 hectares of tobacco without Leccesi sharecroppers, and was hard pushed to harvest the leaves. He and his wife did most of the work, beginning at five in the morning and on some days not finishing until after nine at night – their only 'rest' from the back-breaking work being the three or four hours required to thread the leaves they had picked.

Matteo's family land, acquired from his mother by inheritance, is 0.66.89 hectares; it is mostly arable, but he has a few orange trees, olives and some figs. In 1965 he bought 1.43.48 hectares of arable about an hour away from his house. This was all his property. In addition he rents two hectares of arable in the valley below Caporotondo and plants this with grain. The land is three hours away with a laden cart, and the rent is 5 *tomoli*[1] of grain a year. This is land which his family has rented for thirty years. Matteo also has half a hectare of land which he sharecrops; the owner is an old lady without children or a husband. For twelve years Matteo had a contract by which she provided the land, half the seed and half the cost of reaping the grain, in return for half the crop. Matteo was responsible for the work – ploughing, sowing, threshing and for such fertilizer and insecticides as he used, probably none. After twelve years the widow thought she would get better terms elsewhere, and denounced him to the police for arbitrary possession of the land. The case never came to court, and after harvest Matteo demanded more favourable terms. The land remained uncultivated for three years, after which Matteo took it up again on the terms which he had asked: in addition to paying half the cost of the seed and the reaping, the landlord was to pay half the cost of the string for binding the reaped grain, and half the cost of the fertilizer. In 1965 the land yielded 3 *tomoli* of grain for every *tomolo* sown; this was a very poor yield, compared with the $1 : 5-5\frac{1}{2}$ on Matteo's valley land, $1 : 14-15$ on some other fields at Caporotondo, and $1 : 34$ on a capitalistic mechanized farm on the Metaponto plain.

Matteo used two other pieces of land. Adjoining his own land

[1] *Tomolo* as a measure of capacity is equivalent to about 45 litres. It is also a measure of area and equals about 0.40.00 hectares – roughly one acre. Matteo thus rented 5 *tomoli* of land for 5 *tomoli* of grain, but this one for one correspondence is a coincidence.

there is a plot owned by two brothers who have been in New York for twenty years. Matteo grows grain there, and in return mulches the soil round the few olive trees; the olives go to the owners' mother. For three years he has had other land in the valley for which he pays no rent. Its soil is very poor, and in 1965 the crop of chick-peas turned out so badly that he did not bother to harvest it.

In all Matteo cultivated between four and four and a half hectares of land, to which he has now added the 1.43.48 hectares that he bought in 1965. Of the four hectares, 0.66.89 were his own property.

It was worth Matteo's while to take on very bad land, as long as the cost of cultivation was low. He was so poor that any additional income was worth extra effort, provided it did not require expenditure on working capital. Even the petty outcome of his three-year quarrel with the widow represented a substantial victory for him, since it reduced his costs. Matteo is representative of the twenty-five to thirty extremely poor cultivating families at Caporotondo, who take on sharecropping contracts because they minimize risk. It should be remembered, however, that although Matteo was fairly old – he was fifty-three in 1965 – he is at a relatively early stage in his life cycle, and in fact was just beginning to accumulate the land he would later disburse to endow his children's marriages. His wife was very conscious of the fact that he lagged behind his contemporaries, and railed against her mother-in-law, who had refused to divide her property during her life-time.

THE FANUZZI FAMILY

I end this selection of representative types with a group holding, that of the Fanuzzi family, which can be dealt with briefly. The land is 1.36.05 hectares of arable, and is owned by a mother (who has half the property and a life interest in the other half), nine siblings who have one-twentieth each, and the four children of a deceased tenth sibling who have one-eightieth each. They have never been a farming family; the father was a railwayman who bought the land in common property with his wife for security in 1925 or so, and it has been rented out ever since. At least two of the sisters (owning one-twentieth each) have never seen the land, although their mother had been there. They, nevertheless, regard it as a part of their claim to gentility, and scrupulously register each change of rights of ownership. This is the largest sharing group at Caporotondo; there are half a dozen others slightly smaller, and many families which do not cultivate their land. The smallest sub-division of rights within a single group holding I came across was into 270ths.

The Fanuzzis do not need the rent, which is in fact rarely paid; they talk about their land as 'our small property' and refuse to sell it. The men of the group are on the verge of the professional classes, socially mobile through education and pretension. The women are very conscious of the fact that they have never had anything to do with the land, and perhaps exaggerate their ignorance of it, as they do their ownership. Here we enter the comedy of gentility, where property rights are role performance rights, and not a claim to use land to grow food.

II

People may make a great variety of arrangements about their land. Vincenzo Capece has spent thirty-five years establishing what amounts to a minor territory hegemony, using the income from distant rented land to buy more land close to home, and using his own and his son's marriages to concentrate property within a particular area. Pietro Di Marsico has spent thirty years setting up his daughter, and, having found himself a good son-in-law, aims to eat well, to be generous but not to struggle. Vincenzo Di Marsico, for all his reputation as a property-grabber, has given his land to his children without retaining a life-interest, and continues to work very hard, using his wife's and other land which he rents and sharecrops. Matteo Laviola struggles to find an associate. Abandoned by his brother who has rich affines, he has occasional help from his wife's sister's husband, but this is not enough to enable him to reduce his own labour. Any additional income is worth while, however hard he works, and his tactic is to cut costs as far as possible; he quarrels about the cost of string, and regards the outcome as a substantial victory. The Fanuzzi family uses its land in a purely symbolic way.

There are, then, various alternative arrangements, all of which are generally understood, though they are not equally approved.

Long-term tenancies and sharecropping arrangements for small pieces of land are never between people who define themselves as close kin. People who make land available to others for long periods are either emigrants or the socially and occupationally mobile. Such arrangements are common: all the men in the examples, except Pietro Di Marsico, had held land for periods of twelve years or more, and arrangements which last as long as this may become so fixed that the positions of landlord and tenant or sharecropper become hereditary. This is partly because, although there is a continuing demand for land, any particular piece is more conveniently farmed by some people than by others. The criteria for determining convenience are mainly territorial, and the people who might find it convenient are those who live on or farm neighbouring plots of land. These people, moreover, define themselves as friends if not as kin and are unlikely to act against each other's interests while relations between them are good. Apart from the unpleasantness of quarrelling, neighbours can let chickens stray, pollute wells, and are in good position for campaigns of petty thievery. If anyone had stepped in to take over Laviola's sharecropping contract he would, in fact, have been much annoyed. We might say that the land remained uncultivated for three years, and Matteo won his battle, because none of his neighbours was prepared to become his enemy. And that once established, a tenancy is maintained as long as the relations between possible tenants remain good. Caporotondo landlords are rarely in a very strong position. Their only potential tenants are the cultivators of neighbouring plots, and in my experience they have to bear with delays in paying rent, with the excuse that the harvest was poor, or that the land was not actually cultivated at all that year, with exaggerated claims of expenditure from sharecropping tenants and so on. The other main type of long-term arrangement is informal cession by a group of kin to one of their number. Usually, as in the case of Giuseppe Di Marsico, the beneficiary of the arrangement is the man who would have the land if some formal arrangement were reached; it is related, therefore, to the rights and needs of the cultivator. There are three main types of short-term contract. Pietro Di Marsico's three-year tenancy of a thirty hectare farm is an example of a fairly general practice. Most of the fairly rich peasants at Caporotondo have been tenants of land in this way at some time in their lives, usually when they are preparing for some major expenditure such as marriage.

Pietro's landlord was a successful builder, who had invested some of his profits in land and let it out until he eventually retired from building. At any one time there are likely to be from five to ten of these medium-sized farms up for rent at rates varying between 100,000 and 250,000 lire. In most cases, the would-be tenant applies to a landowner with whom he has a special relationship, or attempts to create such a relationship. Two other types of short-term contract are of recent origin. Tobacco tenancies I have described, and the other sort is used for labour migrants' lands. Not all labour migrants rent out their land. Some go away during the growing season only, returning for the harvest in June and remaining in Pisticci till the autumn, when they sow their land. Others leave their wives in charge, to employ a labourer and a harvester. The labour migrant may make different arrangements from one year to the next.

III

All the cultivators in the examples in Section I farm more land than they own. How does this land become available?

Certain categories of landowner are by social definition non-cultivators. Of the 703 owners at Caporotondo, 374 are women; their land goes into the pool of land available for use by others. About two-thirds of the women are members of groups of owners, which I discuss below. The other third – 118 – have property rights which are not shared with anyone. They make up 17 per cent of the landowners, with 29 per cent (122.35.82 hectares) of the land at Caporotondo. Most of this land is cultivated by husbands or sons or, occasionally, sons-in-law, but up to about one-fifth rent out their land or farm it with labour. Other individuals whose land goes into the pool are men who are labour migrants, and men who are socially mobile.

Turning to groups of owners, the situation is summarized in Table 17.

Spouses who hold land together have in most cases bought it during their marriage. Of the twenty-one couples, fifteen had got their land by purchase; in one other case two husbands of sisters bought land and then divided it. One couple got their land by gift, and four had held it since 1930, so that it is not possible to say how they acquired it. In most cases such joint holdings are of short duration: four couples had had their land before 1930, three before 1940, three before 1950. Two bought land between 1950 and 1954, five between 1955 and 1959, seven between 1960 and 1962. This does not indicate that more couples are buying land in recent years, but that the turnover is fairly rapid – much more so than in the case of inheritance. This is because land is more often bought in order to endow a forthcoming marriage than as a permanent addition to a patrimony.

TABLE 17: Caporotondo: Property-owning groups and amount of land by relationship of members, 1962

Relationship	No. of groups	Amount of land	No. of persons
Spouses	21	19.79.14	42
Parent(s) and child	37	39.54.94	80
Parent(s) and children	38	45.35.55	182
Siblings	20	21.42.45	68
Other	28	25.48.07	105
Total	144	151.60.15	477

Thirty-seven groups comprise parents and one child. Sixteen of these were created when the parents gave property to children; usually the parent gives the land and retains a life interest (thirteen cases out of sixteen). Two were established by marriage contract. Eight further groups, established by supplementary conveyances, were originally established by gift or marriage contract. The total for gifts and marriage contracts is, therefore, twenty-six; for inheritance it is nine; the two others date from before 1930.

Neither of these types of group makes land available for tenancies or sharecropping. The thirty-eight groups of parents and more than one child present quite a different picture. In many cases the Cadaster is out of date; the parents may be dead and in some cases the children are too, although no change of ownership has been registered. The land itself is usually the parental reserve: what the parents keep when they have disbursed most of their property to individual children. Twenty of the groups were established by inheritance, eleven by gift or marriage contract, three were bought, and three date from before 1930. For one I have no information. In eight of the eleven groups established by gift or marriage contract what had happened was that one of a group of co-heirs had given his or her rights to someone else, either as a gift or in marriage contract. The 'gift' is therefore, in these cases, a supplementary conveyance, like a cession following an inheritance: typically, this land is actually farmed by one member of the group.

Eight of the twenty sibling groups originate in a joint inheritance, four in gifts, two by purchase. All four of the gifts were made to groups of sisters. Six groups have lasted since before 1930; four since the 1940s, seven since the 1950s and three were established after 1960. Sibling groups thus tend to last rather longer than others, partly because the members are near to each other in age: where groups include members of different generations the membership changes as they die or transmit their property. Sibling groups last longer, too, because the members of the group have to agree to divide the property, and in some cases this takes a long time. In each case, the land was farmed by one member of the group, or by one other person: there were no sibling groups which farmed the land together. There were a few parent(s) and children groups (e.g. Vincenzo Capece's) where the authority of the parent was strong enough to make the siblings work together; but it was commoner for the land to be ceded informally to one member of the group. There were thus seventy-six groups at Caporotondo, of parents and children, siblings and others (e.g. cousins). In most of these cases the land was available for informal cession to one member, or to a third party, or for renting out. About ninety-two hectares, or 21 per cent of all the land, were affected.

IV

In the last century some landowners have become rich, others have become poorer. In this section I consider briefly how this came about, and then how many people have changed occupation.

The desire for upward mobility is to all intents and purposes universal; at least, this is a safe first assumption for Pisticcese peasants and labourers. Opportunities for them to become mobile, though still limited, have increased: commerce and administration are expanding sources of highly valued employment; and the indigenous 'industries', although by no means secure, offer jobs which are valued because of the fixed hours and the monthly income, and because women like their husbands to have jobs which it is impossible for them to share.

The commonest and simplest way to acquire wealth is to rent land, to produce surpluses for sale, and to buy more land with the proceeds. The secure base of ownership is then used to support a long-term venture, such as educating a son or financing a shop. This is sometimes an alternative to endowing marriages.

There are other sources of income from the land. Some men are reputed to have acquired land with the proceeds of brick-making, using the clay from the barren slopes of the hills – the *calanchi*. Charcoal is still used for heating in town, and some men are said to have made enough money to buy land by charring the prunings from their olive trees and selling them for fuel. All Pisticcesi have the right to collect certain wild fruit anywhere in the Pisticci countryside subject to commonsense limitations, by no means always observed, about standing crops. This right is used nowadays only by small boys and by the unemployed as a stand-by source of food.[2] But at Caporotondo there was one man who, forty years before, had earned enough money from charcoal and *lumpasciòn* to buy three plots. The whole family, he said, had worked themselves to the bone to do this. There were other men, not at Caporotondo, of whom the same was said.

Some people acquire wealth by marrying girls richer than they are. This is unusual; most such marriages are bitterly opposed by the girl's parents, and if the couple elope the parents are likely not to endow the marriage. There are rarely more than two or three elopements a year, out of about a hundred marriages. About 1947 a man at Caporotondo eloped with the daughter of a rich peasant. Her parents refused to recognize the marriage until ten years or so had elapsed, when they endowed the couple with land, paid for the education of the son of the marriage, and settled land on the daughter. But for the most part wealth is acquired through the markets: by selling grain and olives, charcoal, and, formerly, wild fruits, and by buying land.

Education is expensive. From the age of twelve a child's school books cost about 30,000 lire a year – equivalent to four months'

[2] The wild fruits are asparagus and *lumpasciòn*, like an onion but with a softer flesh and a milder taste. *Lumpasciòn* are pickled and considered a delicacy. They are no longer an important source of income to anyone, although the knowledge that they once were was still sufficiently alive in 1965 for the mayor to recommend to the agitating unemployed that they should dig *lumpasciòn*.

old-age pension. From the age of fourteen school fees have to be paid, and children who go to technical schools, teacher training colleges or seminaries have to board away from home or travel every day. University students may have to go to Bari, Naples or Messina for some of their courses. There was one university student at Caporotondo in 1965, and several men who had continued education after fourteen. In addition there were four who had completed secondary education and were employed as clerks in government offices. There were two teachers. All these were the sons of people with rights at Caporotondo, and had been financed with the proceeds of farming at Caporotondo.

Eight or nine shopkeepers with rights at Caporotondo had floated their business with the proceeds of agriculture. One man was in the process of opening a wineshop in 1965. None of the shopkeepers dealt in farm produce. One had a bar-café, one an ironmongery, two were haberdashers, one sold domestic electric machines, a couple were neighbourhood storekeepers selling mostly groceries. There were two artizans, whom I was unable to trace. Two men had jobs in the petrochemical factory.

In addition to these, who can all be said to have been upwardly mobile, more than twenty men were either building labourers or aspirant labourers earning part of their income from the commune's public works office, or from the Forestry Commission. All had some income from the land. Some of these men were in the first difficult and sometimes prolonged stages of becoming occupationally mobile. A further ten men were permanently employed in building, and did not appear to be dependent on the land, although they found time to cultivate it.

A few permanent emigrants have rights at Caporotondo. None of those who lived in Italy had moved into white-collar occupations, but were restaurant waiters, builders or railwaymen. There is a large colony of Pisticcesi at Toronto in Canada, but, as far as I could discover, most of the dozen or so men from Caporotondo were bricklayers or labourers. Caporotondo people who have white-collar occupations tend to stay in Pisticci.

All the white-collar workers had extensive property rights. There were also two families who had become medium land-owners (owning from twenty to fifty hectares). They had holdings at Caporotondo equivalent to three and a half to five and a half 1866 plots. Shopkeepers had varying amounts of property. Some

K

was excellent land with investments made in it after their change of occupation; some consisted of small discrete patches. The two men whom I judged to have the most prosperous shops, selling specialized lines and up-to-date goods, had more than two and a half hectares of good land – the equivalent of two 1866 plots. Both of them had built elaborate houses with cellars and sitting-rooms instead of the more usual all-purpose living-sleeping-cooking-eating room. The all-purpose neighbourhood storekeepers had less good land.

The numbers are too small and the evidence too inexpertly collected for these remarks to indicate how much land a Caporo-tondo peasant must accumulate to become a shopkeeper or clerk. But they are enough to suggest that upward mobility is associated with ownership and not with any of the other ways of getting access to land. And this in turn suggests that upward mobility is associated with a concerted family effort in agriculture to acquire more land either over two or more generations, or during the life of a nuclear family household. Social mobility is not, generally, a matter of individual enterprise and determination. Having a large family may make the individual members poorer, and lead to extreme division of holdings; but a large family is also a large labour force, and the efforts of the whole group may be directed to accumulating wealth and to making one member upwardly mobile.

I was not able to collect exhaustive information about the origins of the present white-collar and professional workers in Pisticci. Although this category is not very large[3] the task required survey techniques which I could not afford to employ. The information is very slight, therefore, and I present it without further comment. Of the forty-two elementary schoolteachers with established posts teaching in Pisticci in 1965, fifteen were descendants of men who had received land in emphyteusis from the communal demesne after 1860. Another two had married into

---

[3] At the 1961 census there were 970 people with service jobs, some of them being manual workers (e.g. bus drivers). They were distributed as follows: commerce, 254; transport and communications, 203; banking and insurance, 23; services, 207; public administration, 283. Another census table, showing the position of all workers in their occupation, gives the following figures for entrepreneurs, professional workers, executive and administrative grades and clerks. Agriculture, 19; industry, 18; all others, 324. This last includes most shopkeepers. ISTAT (1964) Tav. 6, 7 and introductory notes.

such families, indicating that the families had been upwardly mobile. Of twenty-seven shopkeepers on the main street, five or six were descendants of holders in emphyteusis; of three clerks in the Cadaster office, one; of four or five in the Legal Registry, one or two; of four priests, one; of seven party and labour syndicate secretaries, two.

V

Land becomes available for cultivation by another person when the owner is a woman, when he is socially mobile, when the ownership is shared by a group. It can be ceded informally to someone who cultivates it; it can be rented, sharecropped or

TABLE 18: Caporotondo: By what right does the cultivator cultivate?

	No. of properties	Amount of land ha.	%
He is:			
Owner[a]	70*	38.81.48	34
Husband of owner	21	16.30.57	14
Labourer employed by owner	22	10.99.21	10
Son-in-law of owner	1	1.13.08	1
Member of group[b]	20	12.45.12	11
Kin to group[b]	4	3.39.54	3
Renting tenant	18	15.90.24	14
Sharecropping tenant	11	8.12.42	7
Unknown	9†	6.25.00	6
Total	176	113.36.66	100

* Includes 4 of unknown size.   † Includes 2 of unknown size.
  [a] I have included here all land owned by spouses jointly, and farmed by the husband.
  [b] The cultivator has exclusive rights to cultivate, obtained by informal cession.

farmed by labourers; it can be farmed by a husband or son or son-in-law. The Pisticcese term for the informal arrangements is *bonariamente*, good naturedly.

I began to make a census at Caporotondo, asking who farmed the land and by what right it was farmed. The practical difficulty of doing this when people do not live on their land, and when their holdings are scattered, was considerable; and I found that these two questions aroused intense suspicion, even among people well-disposed enough not to think of me as a tax-collector. I therefore abandoned the census when I had covered about one-quarter of the area of Caporotondo. The results are summarized in the following table and map.

There are forty-nine cultivators who farm all the land shown in the map; they own slightly more than one-third of the land

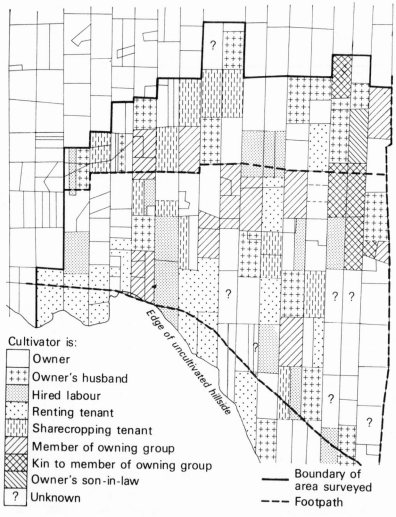

Cultivator is:

☐	Owner
+++	Owner's husband
⋰	Hired labour
⋯	Renting tenant
⫶⫶	Sharecropping tenant
⫽	Member of owning group
⊠	Kin to member of owning group
╲	Owner's son-in-law
?	Unknown

— Boundary of area surveyed

--- Footpath

Edge of uncultivated hillside

Part of Caporotondo: By what right does the cultivator cultivate?

and, through their wives, have secure management of nearly half. A further 10 per cent is farmed by labourers. All the rest (42 per cent) is available for one of the reasons specified in the previous sections. Fifty-one plots are rented, sharecropped or farmed with labour. Fifteen of these are owned by people I define as socially mobile: shopkeepers (three persons, five plots), teachers (two persons, five plots), and people with large holdings elsewhere (three persons, five plots). Eighteen plots (eight rented, three sharecropped, seven farmed with labour) belong to old women. And another four belong to one old man with a U.S. Army pension. Eight plots belong to emigrants.

The last chapter began by showing how the number of owners has increased and the size of holdings diminished over the last century. A closer look at these figures revealed that the range of holdings from the average size has also increased, and that, taking scattered holdings outside Caporotondo into account, the owners there have rights of ownership to 21 per cent of all peasant hold-ings in Pisticci. Most transactions in land take place between living kinsmen, and land is held for periods which, by the standard of our stereotype of peasants, are quite often very brief: between ten and fifteen years.

This chapter has discussed other arrangements for access to land. In the first section, I show that in most cases owner-cultivators cultivate much more land than they own. And the last three sections have been concerned with the way the land becomes available. Many of the contracts, and all the informal cessions, are between kinsmen; and people choose their spouses partly because they have land near their own. I therefore end this chapter with an account of some kinship and affinity relationships at Caporo-tondo.

VI

I have been anxious to avoid giving the impression that kinship in Pisticci society is a peasant equivalent of a sociometric survey diagram. It would be quite easy to describe Pisticcese kinship and affinity as though the lines one draws in making a chart repre-sented frequency of interaction, leadership, reputation – as though, in short, the Pisticcese were children in a school class, who create a group structure after being thrown together by accident.

In the town, we have seen, there are local groups of *familiari* selected from among the qualifying persons in a way which we perceive as patterned and regular. The principles of selection are locality and 'good relations', the latter depending on rules for the transmission of property and the negotiation of marriage settlements. What happens in the country?

The first point to notice is that interaction is much less intense in the country. Houses are more scattered, with at least twenty yards between them; there are places in the country where one can achieve solitude. Most people visit the country, but do not live there except at busy times; those who do live in the country are either past the age of political and sexual activity, or are people of no account, who have no political standing and no honour. There is no neighbourhood community of women to maintain reciprocal control on behaviour, and for this reason not many women go to the country, except at privileged times.

The second point is that the rules for the transmission of town property (houses) and of country property (land and houses or shacks) are different. While women get town property, both men and women get country property. Some men whose patrimonies are scattered may attempt to divide them so that their sons are territorially separated; but unless there are roughly equal amounts of good land in each division, the brothers are likely to resist that. It may be said, for example, that Vincenzo Di Marsico resisted the various divisions and cessions proposed to him because his brother Pietro got better land, even though it was a smaller area. Some marriages are arranged (as were Vincenzo Di Marsico's, Antonio Capece's and Vincenzo Capece's) to consolidate landholdings in a particular area and these, too, affect the composition of local groups. We expect to find brothers and sisters with neighbouring fields; we may find the same with husbands and wives. In some cases we find that cousins marry in order to consolidate their holdings; in this way, holdings which were divided in one generation are brought together in a subsequent one.

There are many differences between houses and land, but what is relevant for our purpose is that houses are not divisible, and land is; and conversely, houses are not summable, and land is. The first is a matter of social rules. The rule is that houses are built with one bedroom, and that there should be only one sexually active couple in a bedroom. The consequence is that a house, even if it is the shared property of several people, is the residence of only one couple. Land, on the other hand, may be, and often is, divided between its owners, and worked independently by them. So, too, while some Pisticcesi have two residences, one in the town and one in the country – and may belong to different sorts of household in each case – nobody has three. If a couple have two town houses, they rent one out until they can give it away with a daughter. If a couple have two country houses they use one and rent out the other. One couple, one residence – or at most two. Land, on the other hand, can be added to without theoretical limit, the more the better – though, if there is a really large accumulation, it is wise to consolidate it. The consequence is that while there has to be a rule of residence – a man marries into his wife's town neighbourhood – there is no corresponding 'rule of work'. A man works where he can, and is not automatically with his wife's sisters' husbands, or his brothers, and so on.

These three points have as consequences that the configuration of kinship in the Pisticci countryside is different from that of the town. The rest of this chapter specifies the characteristics of country kinship; and in the last chapter I contrast these with town kinship.

I found it quite difficult to collect systematic information about kinship and affinity at Caporotondo. It was easy to collect notes from time to time: Giuseppe's visitor was his mother's brother's son, and so on. People, however, were sometimes reluctant to give the names of all those whom they recognized as kin; and I frequently found that the genealogy I thought I had gathered was actually a list of neighbours to whom links were traced through kinship and affinity. So, some Caporotondo landowners whom I visited in town would give me a list of kin which differed widely from that which, in the country, they had assured me was complete.

It is possible to trace links of kinship between all but one of the Caporotondo landowners; and all but six can be linked to the others tracing the relationship only through people who have or had rights there. Such a configuration of kinship has specific characteristics. Each person, I have said, can trace his relation to most other people; he may not be able to trace the links between other people, except by tracing their relationship to himself: when they recall their relationship, however, he may not appear at all. A man is more likely to know the direct relationship between people whom he sees frequently, than between people he sees rarely. Similarly, a man who is asked to trace his relationship with another will consider which of his close associates is linked to that person: he looks first in his own circle to establish that – say – his sister's husband is a cousin to the third party. Consequently it can happen that two men who know they are related, trace their relationship through different links. Another cause of the same

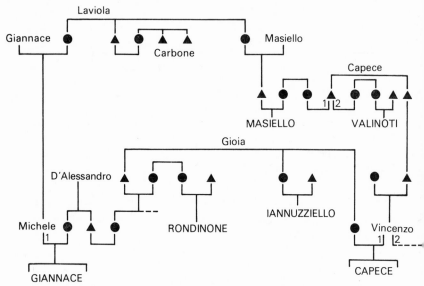

FIGURE 3. Relationship of Capece and Giannace families

phenomenon is that kin quarrel; a line of relationship which one party traces can include people who are thought of by the other only with pain or anger. So, the Giannaces trace their kinship to the Capeces through their paternal grandmother, whose sister's son's wife's sister married a Capece (see Figure 3). But this line includes the Masiello-Capece marriage, resented bitterly by Vincenzo Capece (see above, p. 123). The Capece household thus trace their relationship to the Giannaces through Vincenzo's first wife, a Gioia, whose brother's daughter's husband's sister married Michele Giannace. From the Capece point of view the link through the Laviolas might also be difficult to accept because the Laviolas are close to the Carbones and there is long-standing bitterness between the Capeces and the Carbones over a sexual misdemeanour some decades ago. Finally the Gioia line includes the prestigeful Rondinone and Iannuzziello families.

Another characteristic of Caporotondo kinship is that in the more densely settled areas people include more generations in their kinship reckoning: so, in the relatively densely and permanently settled south-western part of Caporotondo people regularly traced relationships between living adults through four generations. In the eastern part, however, where holdings are larger, and fewer people live regularly in the country, the links are traced through two or at most three generations. It is reasonable to suggest that this variation is the consequence of the greater complexity of relations in the more densely settled area.

Kinship and affinity at Caporotondo is ego-centred: people know the links between themselves and most others, but do not necessarily know the links between others. Two men may use different links to trace their relationship; and the generation depth of kinship reckoning may vary with density and permanence of settlement.

Some cousins' marriages are associated with second marriages. There are six second marriages at Caporotondo, two of which appear to have had little consequence for property (e.g. Pietro Capece's marriages, above, p. 123). In another, a widow with no children, was remarried to her deceased husband's sister's son, in the hope that her property would be secured to her future progeny: she had none, and when she died her property passed to her siblings' children. The other three second marriages involve really quite ingenious property transactions. So, Francesco D'Alessandro remarried a childless widow when his first wife, a cousin, died, leaving two sons. His second wife had some land, and her sister had married her husband's brother. As will be seen from the chart (Figure 4), Francesco's second wife's sister's daughter stood to inherit land from his second wife and her first husband; from her own mother and from her father. She was thus the heir to four people, for about three hectares. So, Francesco's second marriage was coupled with his son's marriage to the heiress, his stepmother's niece. There is in consequence, a progressive assimilation of land into the D'Alessandro patrimony: Francesco acquires the use of his second wife's land, and her first husband's; Antonio, his son, has the use of his wife's land, and when his stepmother dies, he acquires – through his wife – use of the land currently used by his father. Antonio's children will eventually own all this land.

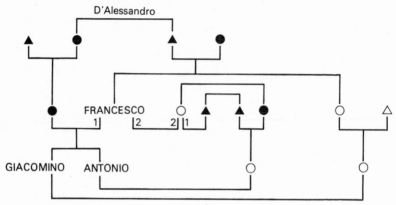

FIGURE 4. Marriage strategies: D'Alessandro

A similar pattern may be observed in another case (Figure 5). Michele Giannace remarried a spinster who was to remain childless. Her heir was her sister's daughter, who was promptly married to Michele's second son, whose children own the marriage settlements of both sisters. Finally, Vincenzo Capece's second marriage

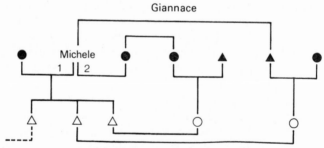

FIGURE 5. Marriage strategies: Giannace

also introduced an heiress into his family (Figure 1). These three families also provide us with four of the nine cousin marriages: Francesco D'Alessandro's first wife was his father's sister's daughter; and his son Giacomino married Francesco's sister's daughter. Michele Giannace's son married his father's brother's daughter; and Antonio Capece married his father's father's brother's son's daughter. All of these reunite property which had been divided in earlier generations, and prevent it from passing 'out of the name', or recover land which has already done so. Cousin marriages on their own do not perhaps allow us to come to any firm conclusions about patterns of marriage and the transmission of property. But the association of cousin marriage with re-marriage and a coupled marriage in the next descending generation does seem to permit us to argue that incisive and ingenious strategies are devised and put into action, to manipulate kinship obligations through marriage, and to create small territorial hegemonies.

It should be emphasized that all Caporotondo marriages involve some property transfers; cousin marriages may consolidate property within a territory; re-marriage *plus* next-generation marriages increase the property available.

To summarize, most farmers who have taxable rights at Caporotondo cultivate more land than they own. This land is made available by kinsmen, by friends, by emigrants, by the socially mobile, and by women. In the course of a normal domestic cycle, the amount of land a man has access to increases and then shrinks. By marriage and by formal and informal use rights a man is able to counteract the rules which lead to increasing subdivision and scattering of holdings. Of course, this is an 'irrational' system: the precariousness of informal rights, the relatively short duration of a tenure, make investment unwise; the settlement pattern makes intensive cultivation impossible for the majority; the concern for the morality of women restricts the amount of labour put into the holdings. Nevertheless it is a 'system' controlled and manipulated by the peasants; it is their own creation, not that of the impersonal working of inheritance laws. Irrational it may be in the eyes of an agricultural economist, but it has its own coherence, and its own consistency with the other systems which we arbitrarily distinguish: with the piecemeal and 'irrational' diversification of the local economy and its consequent encouragement of part-time farming, and with the intrusion of a particularistic bureaucracy and political system and its consequent incentive to town-dwelling.

9

Some Effects of Modernization
Programmes

Two development programmes have affected Pisticci since the war. In 1950 the Agrarian Reform laws brought about the expropriation of 10 per cent of Pisticci's farmland and its redistribution to poor people in lots of up to seven hectares. Pisticci was affected by the national programme because some of the large estates on the Metaponto plain fell within the categories made liable to expropriation. In 1959 methane gas was discovered in the hills below Pisticci, and a factory was built to exploit it. This is part of a *nucleo di industrializzazione* which extends along the Basento valley into the territory of Ferrandina, and it began operations in 1964. In what follows I concentrate on the effects of these programmes within the town. They are very difficult to isolate; Pisticci has become more involved in the national society, more prosperous and more modern as a result of trends and changes which have nothing to do with specific programmes of development. I have focused my account on the limited effects which are directly attributable to these.

I

THE AGRARIAN REFORM

In the years following the collapse of the fascist state, Pisticci, like other towns of the south, went through a period of workers' agitation, occupation of land by the landless, violent repression by *guardaspalle* ('shoulder-watchers' – clients rewarded by landowners for guarding their interests against the encroaching poor), and frequent labour riots. The communist party attempted to organize the discontent and make it an effective political force, and the government responded with emergency decrees and ultimately with a bill for a general agrarian reform; this bill never passed through parliament, but became the basis for two laws covering limited special areas. These were the *Legge Sila* and the *Legge Stralcio* (pruning, or compromise) of May and October 1950 respectively. Pisticci was subject to the *Legge Stralcio*.

The Agrarian Reform Board acquired 2,280 hectares from large landowners in Pisticci, and resettled 322 landholders who were endowed, by means of a variety of grants and mortgages, with land, a house, plants, tools and livestock, and in some cases with furniture, cooking utensils and clothing. Some of the new farms were later irrigated; all are covered by the extension service, which is tied to a co-operative organization for marketing basic crops (grain, olives) and for providing tractor pools, credit, seed, fertilizers and insecticides. A social work service supported settlers with unusual difficulties, and for many years was primarily concerned with winning votes for the demochristian, government, party.

About half the settlers were given farms with a full complement of capital goods; the others were given 'supplementary' holdings, rather smaller (between one and three hectares, against the six to seven hectare farms) and usually not so well endowed. The supplementary holdings were intended originally for poor smallholders with property of their own which was not enough to support their families. But the demand was so great that many were given to people who had no other land. For these people the supplementary holdings were themselves inadequate, and they have been absorbed into the established part-time economy of Pisticci.

In its emphasis on the family farm and family labour the Agrarian Reform was the descendant of the redistributive programmes of the nineteenth century. It was traditional, too, in the assumption underlying many of the practical innovations, that the only thing which could be done for Pisticci and other similar towns was to create a category of prosperous peasants – to improve the prestige and returns of peasant cultivation relative to other sectors of the national economy. The new element was the heavy capital investment in houses and equipment, and the attempt to break down the settlement pattern and make people live on their land, which they would then have time to cultivate intensively.

The Pisticcesi responded very quickly to the Agrarian Reform. The 322 farms or *quote* were heavily over-subscribed: there were 1,448 applications by 1956, and of these 1,158 were judged to be from people 'presumably qualified' to receive land. Of these, 322 got land in Pisticci and another 127 in Montalbano, a neighbouring town where there was an excess of land over local demand. Five Pisticcesi were employed as technical or social workers in the Reform in 1965; the number had been much higher in 1957. The Reform Board also uses local services in education, health and religion. The Animal Insurance Mutuals employ local veterinaries, the bus service is extended to the resettled areas, the bakers deliver bread, and so on. The Reform was a considerable benefit to the teachers, priests, doctors and school-beadles (among others) in the towns.

The large landowners have benefited too. The irrigation scheme, which was the most important single investment after the houses, was entrusted to the Metaponto Land Improvement Consortium, an association of landowners which included the Reform settlers. The consortium has brought irrigation to the estates as well as to the resettlement areas, and some landowners have been able to create efficient modern farms producing fruit, vegetables and flowers for local and national markets. In a good year, mandarins and oranges ripen between two and three weeks before the crops of the Naples area, and they fetch high prices on the markets.

In 1950 two extremely important events were not foreseen: the Treaty of Rome, and the growth of the Italian economy as a whole. The Agrarian Reform, conceived as a means of improving the lot of hapless labourers with no other opportunities, was threatened by both of these. The Treaty of Rome created the possibility of earning better incomes in the secure employment offered by the industries of northern Europe, and at one time seemed likely to drain the larger part of the labour force from the not yet prosperous Reform farms. The industrial expansion in the north of Italy not only offered similar opportunities for employment, but the greater prosperity which it created filtered through to towns like Pisticci, stimulating demand and raising acceptable standards of living.

THE PETROCHEMICAL INDUSTRY

The discovery of methane gas in the Basento valley was followed by a period of intense political negotiations. Should the gas be piped to the nearby towns for heating and cooking? Should it be processed on the spot? Should it be piped to refineries on the coast? In 1962-4 ANIC built a refinery at Pisticci Scalo. Roads, hotels, a residential village, were also built. The river banks were strengthened and electricity was laid on.

The capital investment was enormous, but the amount remains undisclosed. Pozzi, another firm persuaded to set up a chemicals factory in the valley at Ferrandina, was reputed to have been in crisis by 1967; Montecatini, another chemicals firm, at first agreed, then wavered, and finally decided not to build in the valley at all. Local critics of the scheme asserted that it was uneconomic for the firms. They explained the presence of the two companies by the fact that the most important local politician, Emilio Colombo of Potenza, had been able to put pressure on ANIC because he was minister of the Treasury, and on Pozzi because he is a powerful and devout Catholic – a reference to the alleged Vatican majority shareholding in Pozzi. The critics also raised doubts as to the benefit that the local population might gain from such heavy investment: the time required for an equivalent sum to be put into circulation locally in the form of wages and salaries was said by one official to be 'about seventy to eighty years'. This was commonly held to be an underestimate.

The factory and infrastructure were built partly by local labourers, engaged in correct form through the labour exchange. These were mostly married men with families, selected according to the exchange's order of priority. When the work was completed, ANIC took on a new labour force of people under thirty who had been to school until they were fourteen. The factory works three shifts, seven days a week, and each worker has a minimum forty-eight-hour week. Before the jobs were evaluated in 1965 the lowest standard wages for a semi-skilled girl worker was 60,000 lire a month.

There were 1,275 workers employed in September 1965, of whom 265 (21 per cent) were white-collar workers; 300 were Pisticcesi and of these 126 were women. Most of the skilled and semi-skilled labour was taken on and given paid in-training; this was necessary because there were no adequate technical schools in the locality. By giving *borse* (scholarships) to these workers, the firm was absolved from the requirement that they should be engaged through the labour exchange. Nor are they required to apply to the labour exchange for gatemen and porters, who are defined as *di fiducia*, confidential employees. These two categories, trainees and employees *di fiducia*, constitute 75 per cent of the manual labour force. Another 10 per cent were taken on through the War Invalids' Association to fill ancillary posts. In all 15 per cent were taken on through the labour exchange. Two-thirds of these were unskilled workers, and the rest skilled or semi-skilled men who had got their qualifications in north Italy or Germany.

ANIC took some active steps to secure a suitable labour force: the president of the Artizans' Association in Pisticci was approached for the names of young people who might be suitable, and similar approaches were made to the leaders of Catholic Action. The ANIC selection procedure begins with the sifting of applications. (Up to November 1964, some 6,000 applicants had been summoned for interview and given medical checks and aptitude tests.) Discreet inquiries are then made about the general suitability of those who survive the interviews, and the files on the candidates considered suitable are sent to the ANIC head office in Milan. A worker who is engaged receives his letter of appointment from head office.

The discreet inquiries are made by a special body of six or seven internal security men, some of them ex-carabinieri. One of them, who works in Pisticci, is the son of a town policeman (*Vigile Urbano*). Inquiries cover family and personal history, penal convictions, health, and political activities – all matters which might affect the smooth running of the factory. Pisticcesi, who noticed that communists were not taken on as often as demochristians, recounted with anger, or with amusement and pride, the sometimes very odd situations which arose[1] and were inclined to emphasize the political interests of the factory's personnel staff. But the latter did not in fact distinguish left-wing political activity from other counter-productive tendencies, such as drunkenness, tuberculosis, or sexual immorality.

When I left Pisticci in 1966 it was still too early to say what the long-term effects of the ANIC factory were likely to be. Relatively few Pisticcesi were employed there, and these were not influential leaders of local opinion but young men and women. There was a

[1] The secretary of the demochristian labour syndicate (CISL) spent a week in negotiations to have the application of one of his members reconsidered by the personnel office. One Rocco Martino had been turned down after an investigator's report that he was a communist. The secretary was able to prove that the investigator's information was about another Martino, a communist, who lived in the same street.

certain resentment at the factory's policies. The proposal to set up a special school in the factory for the children of northern employees not only angered the unemployed teachers in Pisticci, but was universally seen as implying that Pisticcesi schools were not considered good enough. The white-collar workers in Pisticci were not considered for white-collar jobs in the factory; one surveyor, for example, was offered a job as a warehouseman, but did not take it. The factory, quite rightly, pointed out that the type of qualifications – as rural architects and surveyors, lawyers and farming experts – which the Pisticci middle-class have, were not relevant to the white-collar jobs it could offer. But the young men, educated in the manner traditional for scions of prestigious families, thought their whole cultural background was being undermined and with it their claims to superior status in the town. Nor were the manual workers, many of whom had returned from abroad to build the factory, more pleased than they. There were scuffles between groups of Pisticcesi unemployed and groups of 'northerners' employed to do manual work in the factory – the 'northerners' being actually Sicilians. There was a series of strikes in May and June 1965, mostly by the unemployed, who thought that, since they had responsibilities of family men and the status of adults, they had more right to jobs than women and boys. The factory canteen was supplied not from the local markets but by wholesalers in Bari and Taranto.

Pisticcesi thought the conditions of employment offered by ANIC were better than anything else available in Pisticci. They applied for jobs literally in thousands. They knew, moreover, that the factory could not employ a local labour force on the more obvious objective criteria (ability to do a specific job) because no one in the town had this. They were not very clear what other objective criteria were used: at one time the company would only take on girls taller than 1·6 metres, because the machines they were to work were dangerous for anyone shorter. This was never explained to the girls, and Pisticcesi were hard put to guess what might be the common characteristics of those accepted or rejected. Men who applied for jobs also tended to think that any qualification was adequate for any job, and did not appreciate the degree to which those offered were specialized and non-substitutable. There were also puzzling delays and obscurities in the company's procedures for engaging labour: why should two people who

L

had been told unofficially that they had passed all the tests receive letters from Milan at different times – at intervals, in some cases, of up to three months? One reason was that the man in Milan who signed the letters was an important official, who signed them only when he had time to spare. Pisticcesi were not aware of this rational explanation.

On the other hand, they did know some things. The first personnel officer for the factory lived for some time in a flat owned by one of Pisticci's parish priests. The employee in charge of labour relations in the factory was a lay president of the Provincial Catholic Action Youth Organisation (GIAC), and was, moreover, to be seen in the entourage of the Minister for the Treasury when he visited the town. Whatever the procedures of the company, the politicians and priests made it clear to the Pisticcesi that they controlled who was taken on. It was not only the local town councillors who gave this impression. Shortly before the local elections in 1964, for example, on the eve of the inauguration of the factory, every person who had been taken on received the following telegram:

Delighted to inform you (that as a result of) my interest (on your behalf) ENI has arranged for your admission to training at ANIC Valbasento kind regards Emilio Colombo Minister for the Treasury.[2]

The local politicians established themselves as intermediaries between the population and the factory. The delays in the procedures for taking on workers were not readily explained by the Pisticcesi who, unaccustomed to the ways of rational industrial bureaucracy, were puzzled and prepared to interpret them in terms of the traditional patron–client distribution of resources. The factory was half an hour's bus ride from the town; and since the service is irregular it takes a morning to visit the factory for five minutes. To get into the factory at all one must have an appointment and a pass. There was no ANIC office in town where people could find out how their application was faring. In effect, it was much easier for an important Pisticcesi to get into

[2] *Lieto communicarle mio interessamento ENI habet disposto sua amissione at corso addestramento presso ANIC Valbasento molti saluti Emilio Colombo Ministro del Tesoro.* Quoted in speech by On. Cataldo, Parlamento, Dec. 1964. The curious Italian is telegraphese.

the factory to obtain the relevant information than it was for the would-be workers themselves.

Many Pisticcesi applied to important people – bishops, priests, doctors, the Agrarian Reform officials, the officials of the demochristian organizations – to use their influence to get jobs. All Pisticcesi who worked in the factory, except two, took out CISL cards.[3] Membership of Catholic Action, I was told by a priest, increased by a half during 1964. In the local elections in 1964 the demochristian party won parity on the town council from the left-wing alliance of communist and proletarian socialists, who had previously had a majority of six seats.

The evidence thus suggests that during the period when the ANIC was taking on its labour force, Pisticcesi began to attach themselves to people who were deemed to control its decisions, and to provide themselves with the right political and religious credentials. In fact, it was scarcely necessary for the politicians to have any influence with the factory at all; it was enough for them to be seen in association with the officials – to rent them houses, to walk down the street with them – and to know the procedures, to convince the Pisticcesi that they had influence. On various occasions I was present at conversations in which a political representative told callers that the party would do its best for such and such individuals, and while he could not guarantee that they would be taken on, since the company had its own rules, he could guarantee that some other individuals would not be. Politicians, armed with lists of candidates who had passed the interview and tests, would hold meetings to discuss the political suitability of the candidates in the presence of people who were well placed to spread the news of the party's power. When jobs were evaluated, and salaries adjusted to reward good workers, one parish priest walked up and down the main street giving advance information to any workers or their parents whom he happened to see: 'Your child has done well', or 'I'm sorry, but there's nothing to be done this time'. The reason he had this information, he told me, was that he was concerned to give spiritual guidance to people who might be depressed or puffed up by the evaluation of their work.

The desire to control, or to appear to control, the taking-on of

---

[3] The elections for the *Commissione Interna* (Works Committee), however, gave the communist CGIL two seats, with 200 votes from card-carrying members of CISL. Membership of CISL increased from two to nearly 1,000 in four months.

workers caused conflicts between the *padrùn*, the men of influence, each of whom sought to appear more influential than the others. The parity of the two main parties on the town council was broken when a communist left his party and became a socialist. The demochristians formed a centre-left coalition with him, and he became deputy-mayor and Assessor of Public Works, the office which arranges relief jobs.[4] Shortly after this I heard him referred to as a man who could mobilize the demochristian party's influence on behalf of his friends; and six months later he was seen on the main street arm-in-arm with ANIC employees, though not important ones.

Two of the parish priests quarrelled severely. One, Don Pietro, was a local man, with family connections with the Agrarian Reform officials and considerable local influence; he was the owner of the flat rented by the personnel officer. The other, Don Bergamo, was a northerner, drafted into Pisticci as a missionary; a new parish had been carved for him out of one of the strongest communist wards of the town.

Don Bergamo founded an association, vaguely connected with Catholic Action, called The Christ the King Workers' Movement (*Movimento Operaio Cristo Rè*). The youthful members of this organization went on missionary trips to the hamlets of Pisticci talking to other young men, commiserating with them on the horrors of living isolated in the country. They spoke of Don Bergamo as the only priest with real influence at ANIC: Don Pietro had blotted his copybook by taking the first month's salary from those he had placed in the factory. Don Pietro, they said, moreover, was using these immoral earnings to build himself a yacht at the Barletta shipyards.[5] Don Pietro countered this accusation – which he referred to as an accusation of simony – by preaching to his Catholic Action groups that it is an error to apply for support to priests of parishes other than the one you live in; he used his hierarchical authority to forbid a Good Friday procession organized by Don Bergamo, and a few weeks later Don Bergamo was transferred to another communist hot-bed some twenty-five miles away.

[4] See above, Chapter 2.
[5] The yacht was in fact a fishing vessel built for a company founded by Agrarian Reform officials and others, including Don Pietro's brother (but not Don Pietro himself); the funds came not from the commission on jobs, but from Ministry of Treasury funds for the revival of the fishing industry.

Examples of this type of conflict, though in a more subdued key, can be multiplied: there were disputes between the intellectuals and the artizans in the demochristian party; there was another between the presidents of the War Veterans' Association and of the War Invalids' Association. That between the secretaries of the various demochristian organizations has already been mentioned.[6]

Conflicts between leaders are conflicts about the control of resources which clients are competing for. The conflicts are generally limited because only certain sorts of leader have access to particular resources:[7] jobs with ANIC involved all the people associated with government, as did the Agrarian Reform plots ten years before. In these circumstances the communists keep quiet, and do not advertise their lack of influence. But when they do have control of resources – Public Works, for example – conflict between rivals for these key patronage posts is acute.

II

This brief anecdotal account of the Agrarian Reform and the ANIC factory is intended only to draw attention to two points.

First, the struggle for these resources stimulates the creation of patron-client relationships. Those who are in competition for the resources endeavour to strengthen, or to create, moral and ritual ties with those who are supposed to control them. The leaders of the community see their position of power and influence undermined by the arrival of a new technical elite, which cares nothing for their hard-won scholastic qualifications and culture, and which appears to regard itself as superior to the local men of power.

One of the ways in which local leaders seek to reinforce their position is by trying to control the new resources, or by appearing to do so. It was noticeable that the demochristian politicians and the priests tried to keep applicants for work away from the factory personnel, either by advising them not to seek information and then offering to get it for them, or by preaching against fraternization.

The stimulation of political activity resulted in competition

---

[6] See above, Chapter 2.

[7] Compare Salvemini's account of the Republican Party at Molfetta, 1890–1904, in Salvemini (1962), pp. 55–65.

between clients and between patrons. Recent discussions of patron-client relationships have emphasized the 'vertical' co-operative nature of this relationship; but the concomitant horizontal relationships of competition and conflict between those who are in the market for new resources and between potential patrons deserve equal emphasis. Clientage is activated intermittently so that one man may use several patrons for different purposes; and it is embedded in other relationships, such as friendship, godparenthood, kinship and affinity. What is continuous and latent in all relationships between similars is competition and the struggle for prestige.

Secondly, both these programmes were heavily over-subscribed: there were nearly 1,500 applications for the Reform farms, coming in equal proportions from labourers, landowning peasants and 'others' – shopkeepers, artizans, barbers, clerks and so on. The applications for jobs at ANIC came from as wide a range of persons, though the proportions are not known. The opportunities offered successively by these two programmes were judged to be far better than anything Pisticcesi had ever known. In a community where smallholdings are farmed extensively, and unemployment and underemployment are so common as to be major determinants of the social structure, the introduction of a new modern resource makes all other opportunities appear worthless by comparison. It is not simply a question of creating 2 or 12 or 20 per cent more jobs to mop up 2 or 12 or 20 per cent unemployment. The Pisticcesi were forced to re-evaluate their occupational structure as a whole; and concluded that the new opportunities were better. Just as Agrarian Reform devalued the position of the peasant cultivator, so the ANIC factory devalued that of the Reform settlers. The effect, therefore, is not to reduce pressure on the labourers or part-time farmers, so that they can work secure from competition and improve their own contractual position, but rather the reverse. They, too, become involved in the competition for the new 'better' resources.

Land Tenure, Kinship and Social Structure

You cannot sue an acre; a boundary dispute is not a dispute with a boundary. The study of property rules in general, and of land tenure in particular, is the study of relationships between people. Such relationships 'about property' are not *sui generis*; they are – to put the matter at its weakest – consistent with relationships about other things and activities. Together, the rules of such relationships, and the patterns which conformity to them creates, form a coherent whole.

For example, Stirling suggests[1] that in Elbaşı a map of the farm-land is also a map of agnatic relationships. His Turkish villagers practise partible inheritance among males, and they do not buy and sell land. Consequently, a map of land rights shows the division of fields between the descendants in the male line of some ancestor. When Leach analysed the fragmentation of irrigated fields in Pul Eliya he was able to argue not only that the rights and duties of parents and children were reflected in the pattern of ownership, but that a seemingly confused and (economically) perverse fragmentation could be understood by reference to a community value or principle of equality: water – of varying amounts from year to year – should be distributed equally to each landowner. In Tikopia, the land map is a map of statuses: orchards and houses with their strip of foreshore are named units, and an individual takes the name of the orchard-house-foreshore he owns,[2] even if he does not live there.[3] The houses, and their topographical distribution on the ground – the way they stand in physical rela-tion to one another – have mythical and religious associations,[4] and an attempt to alter this pattern causes trouble.[5] People who claim to have achieved a different status from the one which their

---

[1] (1965), p. 122. [2] Firth (1957), pp. 82–8. [3] *Ibid.* p. 86.
[4] Firth (1957), pp. 342–8. [5] *Ibid.* p. 59.

birth and group membership bestow upon them try to change the names of the houses they live in.[6] A man who is deprived of his land is deprived of his status, of his social personality, and takes to the woods, eventually to the sea.

Pisticci is a more complex society, I think, than any of these appears to be. Nevertheless, apart from the intrinsic interest of studying land tenure there, the principle of analysis is a useful one; it enables us to emphasize features of social structure which might have received less than their share of attention if I had studied, say, manipulative and often ephemeral political relations. In other words, Pisticcesi have at their disposal a number of associative notions with which they are all familiar and which they can, so to speak, pluck out of their repertoire and apply in various situations: friendship, clientship, citizenship are all latent possibilities, sets of norms which may be invoked *ad hoc*, with varying appropriateness and persuasiveness in different situations. But to concentrate on them, I think, distorts the final description of the social structure.

To begin at the beginning: The territory of Pisticci is about 23,054 hectares, and most Pisticcesi do most of the things they ever do within that territory. It is immediately striking that some activities are territorially segregated: residence, home-life, for the most part, is carried on in about twenty-five hectares, concentrated in one place; agriculture is carried on in the other 23,029 hectares.[7] For the most part women and children – particularly girl children – stay in town, and only men go 'outside' to work. When a woman does this on other than privileged occasions, she is considered disreputable. Those families which reside permanently outside are either Agrarian Reform families, aged families, families of no account, or families which are independent enough, economically and politically, to be able to ignore the judgement of other Pisticcesi that they are of no account.

Property is of different kinds in town and in country: houses in the one, land in the other. These have different qualities: land is divisible and summable; houses (residences) are not. They have different rules of transfer: both are transmitted predominantly at

[6] *Ibid.* p. 83.
[7] In 1961 the official count was that 87 per cent of the population of Pisticci lived in town. The reasons this should be considered an underestimate are given above, p. 2 n. 4.

marriage, but houses go to women, and land to both men and women. The practice is that a man goes to live in his wife's house, which is usually in his wife's natal neighbourhood. If these rules are followed absolutely (and they are followed with sufficient frequency for us to say that the stated rule is observed in practice) a man's marital neighbourhood should include the houses of his wife's sisters, their mother, her sisters and their daughters. There are several such sets of people in a neighbourhood; each male member of each set is related to every other member through their marriages. The relationship between sets is hierarchical. The consequence of the rules for transmission of houses and residence at marriage, and of the pressures created by the hierarchical relationship between sets of *familiari* in a neighbourhood, is that kin and affines in a neighbourhood constitute a local kin group of equals.

In the country the rules for the transmission of property are different: land is divided among sons and daughters, mostly at marriage. Land is summable, and to capture it by marriage is a normal aim and procedure. In order to counteract the consequences of division men seek to marry women who have land near to their own. This may involve cousin marriages in some cases, or the pattern of second marriage *plus* an affinal marriage in the next generation. The consequence is that people with land rights at Caporotondo intermarry; and even within an area as small as 400 hectares there are several territorial concentrations of kin and affines. The people with rights in a territory include – in contrast to the town neighbourhood – brothers, brothers-in-law, and brothers and sisters. These people are all, in theory, equally endowed by division and marriage. So a further contrast between town and country is between town *neighbourhoods*, ranked groups of kin who do not intermarry, and country *territories*, where land rights are shared by intermarrying kin and affines who quarrel when they consider themselves unequal.

We have identified two different configurations of kin – selections of individuals from the range of all possible kinsmen and affines. In each case the chart is a product of rules which say how people who stand in certain status positions – parent, husband, wife, son, daughter – should behave to each other 'about' different sorts of property. In politics in the all-Pisticci arena, kinship is at its most elastic and most stretched, but even there the relationships are prescriptive: they enjoin trust, partiality and loyalty, the conferring of benefits to the exclusion of non-kin, and so on. It is not the prescription which is negotiable, but the ascription – the claim to be a kinsman.

The neighbourhood, the country territory and the political network are linked by honour. The honour of the members of a group of kin in a neighbourhood is a common concern; dishonour is contagious, through women. It is not only a judgement of how women behave, of their sexual morals, but also of how well off they are; poverty and dishonesty are intimately related in Pisticcese ideology. The level of living of a Pisticcese family used to be a function of how much land they owned. Land ownership is not now the only source of a livelihood, but it is still important, so that the honour which is assessed in the neighbourhood is created in the countryside. Moreover, honour is basic coinage in negotiated or stretched kinship. A person who has a few friends and associates who are dishonourable is considered good-natured, but he must have a counterbalancing acquaintance among the honourable. A person who tries to stretch his kinship connections is more likely to succeed the more honourable he is.

We cannot say, I think, that economic relations are basic, and that kinship is merely a descriptive idiom emanating from them; certainly not at the level of the analysis of particular marriage strategies. What we can say is that certain features of the present-day kinship system – the rule that most property should be transmitted at marriages; the rule that daughters should get houses, sons and daughters should get land – are new, not more than 150 years old. They appear also to be coeval with the distribution of land and the diversification of the local economy, with its incorporation into a national market and political system. Since there is no evidence about the process of change in kinship, we are bound to leave that question open to speculation. The lines of speculation are suggested, however, by the possibility that events which are coeval are also concomitant; we look at changes, in the first instance, in the overall distribution of land, and in the local economy as a whole.

If we assume, not unreasonably, that Pisticcesi have an average distribution of practical understanding, we cannot attribute the 'irrational' system they have created to defects in their minds or reasoning. We are therefore led to ask: what could this system be a reasonable response to; or, more bluntly, what were the special features ('defects') of the process by which they have become incorporated and integrated into Italian society?

I have argued that changes in the occupational structure have had two main consequences for the tenurial system. First, people have been able to become socially mobile, and this has released land for cultivation by others, so mitigating the immediate consequences of having smallholdings. Secondly, people have been able to diversify their activities, to become part-time farmers – because the new occupations rarely demand the total commitment of labour. In general terms, the inclusion of Pisticci in the national market system has had the consequence of making plots too small to provide adequate levels of living as these are constantly redefined; while the provision of resources for development by a national government upsets existing hierarchies of occupational prestige. I am inclined to think that all these factors combine to make the fragmentation of holdings a reasonable procedure. The value of land has decreased as new opportunities have been introduced, and people realize that it is possible to get a decent living, owning desirable manufactured goods, without working so hard or in such unpleasant conditions as peasants do, and without using the labour of women. For those who successfully move into the non-agricultural sectors of the economy it is not the economic value of land that is important so much as its symbolic value: the ability to make conveyances to match the various relationships of parent and spouse. For those whose approach to the industrial and service sectors is more tentative, and whose non-agricultural jobs are, as most are, insecure and under-demanding, land is a security: a firm base from which to take risks, and something to fall back on if the venture fails. In each case, land may be fragmented to below levels of viability; but it is the special character of Pisticci's diversification and integration which makes that possible. There is, it is true, a greater tendency for people with the smallest holdings to move out of agriculture, than for people who command larger resources. But it should be remembered that the smallest holdings are not necessarily the product of an inheritance system: some people may lack the strategic skills which others use to counter-weigh its effects; others may suffer the misfortune of having a large number of daughters. In the majority of cases, however, the differentiation of the population into rich and poor peasants is the consequence of buying and selling land.

It is possible to analyse Pisticcese society in terms of the principles on which people negotiate alliances and coalitions; to talk of Pisticcisi as if all they ever do is communicate through networks controlled by entrepreneurs, constituted of friends, quasi-kinsmen, patrons, clients. Such a description would be highly selective. My account is also selective, but I consider that an account of marriage, family, kinship, land tenure is sociologically prior to any account of the universals of exchange, negotiation, manipulation. Certainly, we wish to know how people behave in face-to-face confrontations when they wish to persuade, cajole, overwhelm or even simply to maintain a *status quo*. But we also wish to know why particular people are face-to-face at all: what is it that brings *them* together? Why do *those* two or three have need of each other? Why is it that discussions between families about marriage endowments are concluded more equably than discussions about the same endowments within families? To answer such questions it is necessary to refer to social structure: to the patterns of relationships created by obligations and by sanctions. Rules about land – how it is allocated to different purposes, how it is distributed within the population, and how it is then transmitted – reveal the basic structure of Pisticci society. The application of simple rules about settlement, marriage and property transmission creates the fundamental juxtaposition of town neighbourhood and country territory; of hierarchy and equality. These rules are constraining; but they can be manipulated and exploited, so that people can create petty hegemonies, or improve their position in their neighbourhood hierarchy. The rules can be reinterpreted, as they have been in the last decades as Pisticci's economy has altered, and the context of agriculture has changed. But they are also invented – created – as radically new ways of behaving have been, in the century after 1814.

The Natural Environment of Pisticci

Pisticci lies between two rivers: the Basento to the north, and the Cavone to the south. To the east there is the Ionian Sea, and to the west the town's boundary runs through a narrow and broken plain between a hundred and a hundred and fifty feet above sea level. These boundaries contain an area of ninety square miles. A ridge of flat-topped hills rises gently from the coastal plain in a series of natural terraces. The hills become higher, the plateau narrower and more broken the further inland you go until, twenty-odd miles from the sea, there is little more than a narrow ridge a few hundred yards wide and eleven hundred feet above sea level. It is here, with almost sheer escarpments on three sides, that the town Pisticci is built: the town overlooks the two rivers and a few miles of plain to the west.

The hills are clay, although the highest parts, where Pisticci is built, have a top stratum of sandstone. The clay is extremely liable to erosion and landslides. As surface water drains down the sides of the hills it cuts deep gullies into the clay, strips the hillsides of their topsoil and leaves them bare, grey, razor-edge ridges which slope precipitously down to the valleys below. Because the soil is clay the water does not drain away under the surface, and the earth slides away from under buildings or roads. At Pisticci one quarter (*Rione*) of the town has been evacuated because the houses are endangered in this way; a church is closed, and the main road through the town, the only one capable of carrying heavy traffic, was broken in three places in 1965.

The average temperature on the plain at Metaponto is 17°C, and it is one or two degrees cooler in the town. This represents an average of 8°C in the coldest, and of 27°C in the hottest months. The maximum temperatures recorded are around 40°C. It is typical of the irregularity of the climate as a whole that although the temperature rarely falls below freezing point for long periods, the town was snowed-up twice in eleven months while I was there, in February 1965 and in January 1966.

The average annual rainfall at Pisticci is 748·4 mm.; at Metaponto on the coast it is 603·9 mm. In general the autumn and
winter months are wetter than the spring and summer, which
tend to be very dry.

This is the pattern of the climate. Given the nature of the soil,
which is for the most part impermeable, neither absorbing mois-
ture nor allowing it to drain away satisfactorily, harvests vary
greatly from year to year. In a bad year crops which germinate
during the mild wet winters grow precociously during the early
spring, and then suffer drought and great heat. For spring-planted
crops, such as tobacco and chick-peas or tomatoes, irrigation is

TABLE 19. Metaponto: Seasonal variations in rainfall

Year	Total annual mm.	Winter (Dec–Feb) mm.	%	Autumn (Aug–Oct) mm.	%
1930–31	632	400	63·0	58	9·2
1939–40	677	112	16·8	287	43·0

Source: Kayser, B. (1964, p. 35), who cites an unpublished study by
Rossi-Doria, M.

needed to supply probable natural deficiencies. There was virtually no irrigation until ten years ago; and it is still largely confined to the coastal plain. In these circumstances, small-scale
agriculture is a precarious livelihood in which even subsistence
is annually threatened. Harvests vary between the very poor and,
on occasion, the excellent.

The average figures conceal wide variations from year to year,
and from season to season. As an example of such variation,
Table 19 shows that in 1930–1 nearly 63 per cent of the rainfall
for the year fell in the months December, January and February,
while in 1939–40 only 16·8 per cent of that year's fell in those
months. In contrast, the three months August, September and
October were months of heavy rainfall in 1939, when 43 per cent
of the year's total fell; but in 1930 only 9 per cent of that year's
rain fell in those months.

In addition there is considerable local variation. In 1965, for
example, there was no rain at Caporotondo for nineteen weeks,
from 11 April to 20 August. At Pisticci, five miles away from the

rural Caporotondo area, there were two storms lasting more than
an hour, when the streets were awash, as well as slighter rain on a
number of occasions over a period of three weeks in June. Finally,
when the rain did come to Caporotondo in August, it fell in a
deluge on the southern edge of the plateau, and in a mild drizzle
a mile or so further north.

The Cavone and the Basento are the boundaries of the commune. It is said that the Basento was navigable in the twelfth
century; but now, like the Cavone, it meanders in a wide, shallow,
stony bed. These rivers, as well as the many streams and rivulets
which form, are typically irregular rain rivers, receiving little
water from natural springs. In summer the Cavone dries up completely, and the Basento is at best reduced to a placid stream. In
winter they may be violent torrents, unfordable, carrying away
trees and boulders and the precious topsoil from the flanking hills.
In 1959, after exceptional rains, the Basento river station at
Menzena measured a flow of 1,420 cubic metres of water per
second,[1] and the floodwaters of the Basento, Cavone, Agri and
Sinni inundated the plain between Metaponto and Policoro. The
area devastated by the floods was the best and most fertile land in
the whole province, the area in which the Agrarian Reform Board
and the Land Improvement Consortium had made their heaviest
investment. Clearly, just as the unpredictable climate and the
uncertainty of a steady return have always impeded individual
investment in the south, so now the need to plan for such extremes
of weather greatly increases the cost of state intervention to
control the environment and to remedy its defects. Towns have
to be buttressed because the hills on which they are built are
unstable; the climate is irregular, and there is a perennial danger
of drought in the crucial spring months. 'In these parts', as
Fortunato wrote, 'the sun and the rain never come together.' The
characteristics of the physical environment which are most important are not so much the beauty of the mountains, the heat of the
summer and the gentle mildness of the winters, but rather what
seems a conspiracy of instability, irregularity, and fluctuation
between wild extremes.

[1] Tekne, (1964), p. 44.

The Distribution of Population, March 1963

(i) Pisticci town:

Rione

Terravecchia	579
Loreto	1,058
Dirupo	1,509
Tredici[a]	359
Croci	1,383
Municipio	863
Picchione } Montebello }	1,484
Piro	774
Mattina Sop.	1,026
Mattina Sott.	1,733
Mattina Nuova	695

—— 11,578

(ii) Contrade near Pisticci town:

Contrada

Camarelle	3
Calcarole	3
Coppo Sott.	4
Fondo Messere	6
Fontana del Fico	3
Fontana la Pietra	15
Lagarone	6
Lamia Mozza	4
Masseria Nobile	6
Mesole Cirallo	18
Pantone Ciuchera	5
Paolone	23
Policeto	17
Pozzitello & Rupa	6

[a] The threat of landslide is greatest in this *rione*.

S. Angelo	38
S. Croce	8
Scurafaci	7
Serra Segnata	6
Simone	3
Scalo FFSS	43
S. Francesco	3
S. Leonardo	1
S. Pietro	17
Varcaturo	3
	—— 248

(iii) Marconia district:

Hamlets

Marconia	230
Tinchi	149
Centro Agricolo	108

Contrada

Accio Sott., Sop.	38
Canala	48
Caporotondo	147
Castelluccio	4
Centro Agricolo	94
Demanio	—
Madonna del Carmine	4
Masseria d'Avenia	—
S. Vito	9
Tinchi	316
Madonna del Pantano	11
Rullo	16
Cenapura	10
Feroleto	—
S. Pietro	—
Lavandaio	30
Marconia	332
Olivastreto	6
Serricchio	69
Vomero	59
	—— 1,680

(iv) S. Basilio district:

Contrada		
Centro Casinello[b]	1,251	
S. Basilio Castello	22	
Franchi	10	
Macchia	21	
S. Teodoro	15	
Torretta	5	
	———	1,324
Grand Total		14,830

[b] The chief area of Agrarian Reform resettlement in Pisticci.

APPENDIX III

Some Points of Family Law

The constitution; The penal code; Matrimonial disputes and offences; Matrimonial property; Succession and inheritance; Obligations to give material support 'inter vivos'

THE CONSTITUTION

The family has a unique position in the Italian constitution. It is stated to be a 'natural association' and there are no others; and its rights are recognized. The only other institutions which have their rights recognized are the Republic itself and the religious confessions. The Republic's rights are implicit in the duties laid upon individual citizens. The Church of Rome and other confessions have the right to organize and regulate themselves by their own statutes, which are not limited by the State except in the provision that they shall not conflict with the civil law. Indeed, the Republic and the Church of Rome are recognized as each 'independent and sovereign in its own sphere' (*Cost.* art. 7); and the Lateran Pacts, which regulate relations between the State and the Church, are a treaty between sovereign powers of equal status. The distinction between the family, or religious confessions, and other institutions is one between institutions having to some extent an intrinsic personality, while other institutions have a legal personality and are created or permitted by law. It may be that the reason the family is thus conceptually associated with the Republic and the Church, even though it is not a sovereign power, is that it is a 'natural association' and is seen to have its origin in natural law.

However that may be, other rights in the constitution are 'inviolable rights of man, both as an individual and in the social institutions in which he plays a role (*nelle formazioni sociali dove si svolge la sua personalità*)' (*Cost.* art. 2): and these, whatever the last clause may mean, are clearly the rights of individuals and not of institutions. For example, the law on political parties states: 'All citizens have the right to associate freely in (political) parties to compete by democratic methods to determine national policies.'

The distinction between the family and other non-sovereign social institutions is perhaps best made clear by quoting the law on labour syndicates alongside that on the family.

No obligation can be laid on labour syndicates other than that of registration. . . . It is a condition of registration that the statutes shall establish a democratic internal organization. Registered syndicates have legal personality. (*Cost.* art. 39)

The Republic recognizes the rights of the family as a natural association founded on matrimony. Matrimony is ordered by the moral and legal equality of the spouses, and by the limits established by the law as a guarantee of family unity [or: '. . . of the family unit']. (*Cost.* art. 29)

THE PENAL CODE

Between members of the same family (spouses, ascendants and descendants, adoptive children and parents, siblings living in the same household, and a person and his spouse's ascendants and descendants) certain actions are not punishable, although they are punishable when the persons involved do not stand in one of the relationships listed. These actions are: theft; appropriation of common property; usurpation of immoveable property; deviation of watercourses; occupation of buildings; disturbance of possession with menaces; malicious damage; abusive pasturing of animals; trespass; maiming or killing of animals; defacement of property; fraud; fraud while bankrupt; abuse of legal incapacity of minors or of insane persons; usury; fraudulent incitement to emigration; fraudulent conversion; stealing by finding, and receiving (*Cod. Pen.* art. 649).

Other offences can be extenuated by the fact that the persons involved stand in one of the relationships listed above. For example, it is well known that homicide for motives of honour (defined as the homicide of a spouse, daughter, or sister discovered in an illegitimate sexual relationship, or homicide of the other person involved, or of both) carries a penalty of from three to seven years' imprisonment. And in this case, penalties of less than three years are not uncommon because other extenuating circumstances may be taken into account. Similarly, although causing bodily harm to a member of one's own family carries a slightly higher penalty than causing bodily harm to other persons, the normal penalty is reduced by two-thirds when the offence is 'an abuse of the means of discipline'; that would be, when the victim is the wife or child of the person convicted.

Adultery by wives was punishable until recently on the plea of the husband, with up to one year's imprisonment. Continuing adultery was punishable with up to two years' imprisonment, as was a husband's concubinage (keeping a mistress in the matrimonial home or notoriously elsewhere). The partner in adultery, continuing adultery or concubinage was liable to the same penalties as the offending spouse. A person who abandons the family home or behaves in a way contrary to the good order or morality of the family, and does not maintain his obligations to support the family (see below, *Material support*) can be imprisoned for up to one year. So, too, can the person who spoils or in some other way diminishes the value of property belonging to his minor child, ward or spouse. (*Cod. Pen.* art. 570). A person who falsely registers births or deaths, or who alters a member of his family's civil status (e.g. from legitimate to illegitimate and vice versa) is also punishable.

MATRIMONIAL DISPUTES AND OFFENCES

The civil code lays down the obligations of spouses: the reciprocal obligations of cohabitation, fidelity and assistance; the wifely duty to accompany her husband wherever he may think fit to establish his residence; the husband's duty to protect his wife, to keep her with him, and to give her all that is necessary for life in accordance with his means (but if he has no means the wife is obliged to support him). Matrimonial offences against these norms are grounds for separation: adultery (as the term is now used; formerly adultery by women and concubinage by men), desertion, excess of legitimate power or means of discipline, cruelty, grave menaces, and grave offences to personal dignity (*Cod. Civ.* arts. 143–5, 151). Other grounds are: the imprisonment of a spouse for a period not less than five years; the condemnation of a spouse to perpetual exclusion from public office (*Cod. Civ.* art. 152); the failure of a husband to establish his residence without just cause, or his failure to do so in a way appropriate to his social standing (*conveniente alla sua condizione, Cod Civ.* art. 153). A decree of separation is obtained in a court, which, as a justification for the decree, declares one or both of the spouses to be guilty of a matrimonial offence. The guilty spouse remains subject to all the matrimonial obligations consistent with separation: fidelity, the provision of 'all that is necessary for life', and any property obligations (to make over an income or life-interest, to provide a house, and so on) which may have been entered into at the time of the marriage. (See below, *Matrimonial property*). He or she loses all rights to the sexual fidelity of the innocent spouse, to material support and to any property or income acquired at the time of the marriage. An innocent spouse consequently loses all obligations and retains all rights. In this way, for example, an innocent spouse can sue the guilty one for adultery or concubinage, but is himself immune from action (*Cod. Pen.* art. 561). If both spouses are declared guilty, then each loses all rights, and consequently no obligations lie on either.[1]

MATRIMONIAL PROPERTY

The legislation on family property in the civil code is largely permissive, allowing husband and wife to create special categories of property which are peculiar to the family. Property which comes into one of these categories may be protected against third parties, or it may be subject to special claims by members of the family. This particular status of family property corresponds to the particular status of the family itself in the constitution and in the penal code.

There are four categories of matrimonial property: dowry (*dote*); paraphernalia (*parafernalia*); patrimony (*patrimonio*) and common property (*beni comuni*). The property of either spouse may come into any one of these categories; but it is neither necessary nor, in Pisticci, usual for all property to be classified in this way. For all the categories except paraphernalia a contract must be made and registered by a notary, or the status of the property must be noted on the contract by which it is acquired. The items must be listed and their value stated. Paraphernalia is a residual category: all the wife's property which is neither dowry nor patrimony nor common property must be paraphernalia: no contract is necessary.

[1] In 1970 the law was changed to allow divorce, on restricted grounds. At the time of writing the law was under discussion at the constitutional court.

Dowry is the goods, of any kind, which are given to the husband to support the costs of the marriage. They may be given by any person. Rights of property in cash or moveable goods pass to the husband for the duration of the marriage, unless the contract states otherwise. Rights of property in immoveable goods may also pass to the husband if the contract expressly states so. In either case, the husband is the sole distributor of the goods, and has the sole right to the fruits; he may be required to give security for the goods to the reversionary owner (who may be his wife or a third party). A wife has the right to apply to the courts for administration of the dowry if the husband proves incapable of it or if he administers his own property in such a way as to threaten the dowry. If the wife is the innocent spouse in a judicial separation she may have the dowry restored to her: otherwise whether or not she recovers the dowry rests with the magistrate. Dower goods may not be used as security, nor may they be alienated unless this is authorized by the courts or by the original contract. If it is authorized, the proceeds of the alienation, or of the benefit secured against the dowry, are themselves subject to the ties and regulations of the dowry laws. Dowry returns to the wife or to the donor or to their heirs on the death of either of the spouses. It is protected against all actions by third parties while the marriage lasts (*Cod. Civ.* arts. 177–201).

Patrimony can be either immoveable goods or investments, and property rights in it may belong to either spouse. It must be constituted as patrimony by a contract in which it is stated that the goods are to be devoted to the advantage of the family (as opposed to supporting the costs of the marriage [*matrimonio*] which is the legal purpose of dowry). The administration rests with the property rights. The goods may not be alienated without the consent of a court which must be persuaded that the family is in need, or that the profit is certain. Both the goods and their fruits are protected against third parties; they cannot be seized for debts contracted after they were tied as patrimony. The tie ends when one of the spouses dies, or when the youngest child reaches its majority, whichever is later. This is in contrast to the tie of dowry which ends with the marriage; and the difference corresponds to the different ends to which the property is to be devoted.

Common property can only be the profits and acquisitions made by the spouses during their marriage, or the enjoyment of goods subject to other ties. The administration is in the hands of the husband, who is not permitted to alienate or to use the goods as security if they are owned by both spouses together. The goods are to be used for the expenses of the family and each spouse may claim maintenance from the common goods. They are not protected against third parties: the husband's creditors may take the wife's property to settle debts contracted by him after the marriage, and the wife's creditors may do the same. The purpose of the common ownership is to guarantee to either spouse the funds for managing his affairs, and it gives each spouse a claim on the property of the other. The communion ends with the death of one of the spouses or with their separation. The surviving spouse, and the heirs of the deceased spouse have a right to the property which each originally brought to the marriage, but this right is not protected against creditors.

The category paraphernalia covers all a wife's property which is not in any other category. It is not protected against third parties, and the purpose of it is that it defines the extent of the wife's duty to maintain her husband and to contribute to the upbringing of the children. The value of the parapherns, and the income from the dowry if there is one, may be calculated as a proportion of the total wealth of the family, and the wife can be compelled by a court to contribute that proportion to the family expenditure.

The laws about matrimonial property thus have two effects. On the one hand they permit spouses and others to devote their property to the use of the family, and grant them protection against third parties if they do this with a formal legal contract. On the other hand, a spouse's property is liable to certain claims from the other spouse which no other person may have except, in some cases, their children. The provisions about the administration of goods, and the attribution of property rights, may be seen as reflecting the status of spouses.

SUCCESSION AND INHERITANCE

The laws which regulate the transmission by inheritance or succession of property at the death of owners may be seen as the counterpart of the obligations which lie on owners to support their spouses and children and in certain circumstances other kin (see below, *Material support*) during their lifetime. Although in what follows I speak from time to time of kin being ranked or having equal claim, I do not see these laws as giving a scale of closeness of kin, or a measure of intimacy. It is obvious that the laws do not reflect a homogeneous Italian reality: *a priori* the structure of the family and kinship obligations are not the same in Milan and in Pisticci. An account of the law, therefore, is one of a common legal structure; how far this corresponds to the reality of Pisticci will be clear in the main text.

Property passes at the death of the owner in at least two of three ways: by legitim and will, or by legitim and lawful inheritance, or (if a testator overlooks part of his property) by legitim, will and lawful inheritance. Legitim secures some of the dead person's property to his descendants or ascendants and a usufruct to his spouse; it has absolute priority over all other ways: it represents an overriding duty, and for this reason I speak of succession. A will is the expression of the will of the dead person, who has free disposition of his property except for the legitim. If a person dies intestate his lawful heirs inherit that part of his estate which is not legitim, in the proportions established by law. The rights which may be disposed of are:

to income (by will; or, to some kin, by legal inheritance);
to usufruct, i.e. to an income and to the right to administer the property from which the income is derived (by will; and by legitim and legal inheritance);
to bare property, i.e. to property rights less income or less usufruct (by will, by legitim and by legal inheritance);
to full property (by will, by legitim and by legal inheritance).

Concerning legitim there are three questions which should be discussed: who succeeds, what proportion of the estate is reserved to the successors (*legittimari*), and in what proportions the legitim is divided among them.

A dead person's legitimate children, legitimate ascendants and illegitimate children have claims to full property in the legitim; his surviving spouse has a claim to a usufruct only. A grandchild or a grandparent may succeed in place of the dead person's child or parent; but the claim to a usufruct is not transferable. Legitimate children means legitimized or adopted children as well as those who had both the biological and social elements of filiation at birth. Legitimate ascendants also includes legitimizing or adoptive ascendants. Illegitimate children are those who are recognized by their parent or parents: the filiation is recognized, but this does not create the full status of child nor of parent. As we shall see, an illegitimate child does not rank equal with a legitimate child in succession or inheritance; an illegitimate parent does not acquire rights of usufruct over a child's property during his minority, nor does it establish kinship between the child and the parent's kin (except in special circumstances). Unrecognized children have no rights to the legitim, but have the right to an income from the estate of an intestate parent. It should also be noted that some children are unrecognizable: a child conceived in adultery cannot be recognized by the married parent (or by either, if they are both married) while the marriage lasts; nor can a child conceived in incest be recognized ever, unless the recognizing parent was unaware at the time of conception that the relation was incestuous. A child who is adopted cannot be recognized by the same parents or others. A spouse is excluded from the legitim and from the inheritance if he has been judged the guilty party in separation proceedings, or if both have been judged guilty.

The legitim is secured first to descendants, then to ascendants. If there are legitimate descendants, then ascendants are excluded. If there are illegitimate descendants the legitim is divided between them and legitimate descendants or (in the absence of these) ascendants. The spouse has the right to usufruct. The four categories may thus be seen in relation to each other as excluding or as sharing categories. This is represented in Figure 6 by brackets: the categories which are bracketed together share, those which are not, exclude or are excluded.

Legitimate descendants ⎤ I
Illegitimate descendants ⎥ ⎤ II
Legitimate ascendants · ⎦
Spouse] III

FIG. 6. Legitim: Sharing and excluding categories

The proportion of the estate reserved as legitim varies between one-third and two-thirds according to the category and number of the successors, as will be seen in Table 20. A word should be said about the identification of ascendants and descendants. The

TABLE 20: The Legitim: Its size; how it is divided among successors

Relationship to the dead person	Size of legitim as fraction of estate	How it is divided among the successors*	Art. of Cod.
(A) *Successor from one category only*			
Legitimate descendants			
1 only	$\frac{1}{2}$		537
2 or more	$\frac{2}{3}$	Equal shares	537
Legitimate ascendants	$\frac{1}{3}$	Equal shares	538
Illegitimate descendants			
1 only	$\frac{1}{3}$		539
2 or more	$\frac{1}{2}$	Equal shares	539
Spouse	$\frac{2}{3}$ usufruct		540
(B) *Successors from sharing categories*			
I (i) Legitimate descendant(s) & (ii) Illegitimate descendant(s)	$\frac{2}{3}$	At least $\frac{1}{3}$ must go to (i); otherwise shared in the ratio 2 : 1	541
II (i) Legitimate child & (ii) Spouse (usufruct only)	$\frac{2}{3}$	(i) has full property rights to $\frac{1}{3}$; (ii) has usufruct of $\frac{1}{3}$. (i) then has bare property rights to half (ii)'s usufruct. Testator can dispose of the other half of the bare property	542
III (i) Legitimate children & (ii) Spouse (usufruct only)	$\frac{2}{3}$	(i) divide equally full property in 5/12; (ii) has usufruct in 3/12. (i) share equally the bare property of all the usufruct	542
IV (i) Legitimate child(ren) & (ii) Illegitimate child(ren) & (iii) Spouse (usufruct only)	$\frac{2}{3}$	(iii) has usufruct of 3/12 (i) & (ii) divide full property rights to 5/12 and bare property rights to all the usufruct in the ratio 2 : 1, but $\frac{1}{3}$ is always reserved to (i)	542 & 541
V (i) Illegitimate child & (ii) Spouse (usufruct only)	$\frac{2}{3}$	(ii) has usufruct of 5/12. (i) has full property rights of 3/12 and bare property of one-fifth of the usufruct (i.e. of another 1/12)	543
VI (i) Illegitimate children & (ii) Spouse	$\frac{2}{3}$	(ii) has usufruct of $\frac{1}{3}$. (i) divide equally full property of $\frac{1}{3}$, and bare property of half the usufruct (i.e. another 1/6)	543

* In this column the fractions written with numerals refer to fractions of the estate; the fractions written in words refer to fractions of the legitim.

Relationship to the dead person	Size of legitim as fraction of estate	How it is divided among the successors*	Art. of Cod.
VII (i) Ascendants & (ii) Illegitimate child	$\frac{1}{2}$	(i) to have not less than 1/6, but otherwise share equally with (ii)	545
VIII (i) Ascendants & (ii) Illegitimate children	$\frac{2}{3}$	(i) to have not less than 1/6, but otherwise share equally with (ii)	545
IX (i) Spouse & (ii) Ascendants	$\frac{2}{3}$	(i) has usufruct of 5/12 and testator can dispose of all the bare property; (ii) share equally full property of 3/12	544
X (i) Illegitimate child & (ii) Ascendants & (iii) Spouse	$\frac{2}{3}$	(iii) has usufruct of 10/30; (ii) share equally full property of 6/30; (i) has full property of 5/30 and bare property of 6/30. Testator can dispose of bare property of remaining 4/30 of usufruct.	546
XI (i) Illegitimate children & (ii) Ascendants & (iii) Spouse	$\frac{2}{3}$	(iii) has usufruct of 10/30; (ii) share equally the full property of 5/30; (i) share equally the full property of 5/30 and the bare property of all the usufruct (i.e. of another 10/30)	546

* In this column the fractions written with numerals refer to fractions of the estate; the fractions written in words refer to fractions of the legitim.

share of ascendants does not vary with their number, but according to with whom they share. The portion goes to the nearest ascendant irrespective of line, and if there are more than one equally near they divide in equal shares. But descendants' shares do vary with their number. If the children of a dead person have predeceased him, the children succeed to the share which would have gone to their parent.

Legal Inheritance: the legal heirs to property which is not secured as legitim, nor disposed of by testament, are set out in Figure 7, which may be compared to that showing successors. The brackets, as in the chart of successors, indicate sharing categories of the dead person's kin. I excludes II, II excludes III, and so on. If there are only legitimate children, all others are excluded (except for the spouse who has life interest); but they share with illegitimate children, who (failing legitimate ones) may share with their surviving parent and ascendants of their dead parent. If there are no descendants, the spouse and ascendants share with siblings;

$$
\text{II} \begin{bmatrix} \left.\begin{array}{l} \text{Legitimate descendants} \\ \left[\begin{array}{l} \text{Illegitimate descendants} \end{array}\right. \\ \text{Legitimate ascendants} \\ \\ \left[\begin{array}{l} \text{Spouse} \\ \text{Other kin} - 4° \end{array}\right] \text{IV} \end{array}\right] \text{I} \\ \\ + \text{Siblings} \end{bmatrix} \text{III}
$$

Other kin – 6°] V

Unrecognized and unrecognizable children] VI

FIG. 7. Legal heirs: Sharing and excluding categories

and in default of these two latter categories, the spouse shares with other kin to the fourth degree (first cousins); if there are none of these, then kin to the sixth degree (second cousins) are admitted. Failing all these, the state inherits. Unrecognized and unrecognizable children have the right to an income (i.e. they do not have administration rights over the property assigned for them), but this should not exceed what would be the income from their share were they recognizable and recognized. The widowed spouse has a life interest when sharing with legitimate children, but otherwise receives full property rights.

The definitions of the relationships are the same as for the legitim (see above, pp. 175–6). Siblings of the dead person may be either full or half-siblings: half-siblings are admitted to the inheritance even if there are full siblings; but they receive half the share of full siblings.

In Table 21 there are various possibilities of inheritance which I have not included. Siblings are not the lawful heirs of illegitimate children. Consequently, if an illegitimate child dies leaving neither descendants nor spouse, his estate passes absolutely to the parent (or ascendants of the parent) who recognized him; in default of ascendants or descendants the estate passes to the spouse. An illegitimate child may inherit from his recognizing illegitimate parent's legitimate parent if there are no other kin within the third degree, nor a spouse. These are the exceptions to the general rule that illegitimacy does not create kinship between the recognized child and the kin of the parent.

One other comment remains to be made. In most cases the estate is inherited by categories of persons, who then divide their share equally between them. In some cases, however, all the individuals from one or more categories may share *per capita*. Although I am not concerned here with the assumptions underlying the law about families, we may note that where there is inter-category sharing on a *per capita* basis, the individuals concerned are all one-time members of the dead person's nuclear family of origin (see, in Table 21, the ranks VII and IX, and compare VIII and X where parents are replaced by ascendants).

We may also note the law of legal inheritance places much more emphasis on the nuclear family than does the law of legitim, which, by excluding siblings and other collateral kin, emphasizes the direct line of descent and ascent. Even in the legal inheritance law, however, priority is given to the descendants and then to ascendants over spouse and siblings.

The method of calculating what is to be the legitim of an estate is laid down in articles 556 and 747–50 of the civil code. To determine the quota which the dead man may dispose of by will, all the goods of which he dies possessed are valued and the values summed. Debts are subtracted from this. Then the gifts which he may have made during his lifetime are, by legal fiction, added to the total. The value of gifts is calculated as it was at the time or death, depreciation being allowed, as well as appreciations resulting from the inherent nature of the gift; no allowance is made for depreciation resulting from any damage caused by the beneficiary. The legitim is then calculated as between one and two-thirds of the grand total. If the dead person has disposed of more property than he is entitled to, either by will or by gifts made *inter vivos*, then a successor may claim for the restoration of gifts or the annulment of the dispositions of the will. The legacies made in the will are docked before the gifts (*Cod. Civ.* arts. 554–5). These rules apply only when there is an alienation of property rights in excess of the legitim. The legitim is also protected against lawful heirs: a dead person's legitim cannot be encroached on by his lawful heirs (*Cod. Civ.* art. 553): it is hard, however, to invent a circumstance in which this would apply.

The rights of successors to the legitim are also protected against donations of usufruct or income in excess of the free quota. If the dead person has made legacies of these which encroach on the full property rights of the successors, they must choose whether to fulfil the dispositions of the testator, or whether to relinquish to the legatee the bare property rights which they might have (as

TABLE 21: Lawful heirs: How the estate is divided among them

Relationship to the dead person	Ratio for the division between them	Quota reserved to any category	Art. of Cod. Civ.
I (i) Legitimate descendants^a & (ii) Illegitimate descendants	2 : 1	(i) $-\frac{1}{3}$^b	574
II (i) Illegitimate descendants & (ii) Legitimate ascendants	2 : 1		575
III (i) Illegitimate descendants & (ii) Spouse	2 : 1		575
IV (i) Spouse & (ii) Ascendants	1 : 1		582
V (i) Illegitimate Descendants & (ii) Spouse & (iii) Ascendants	5 : 4 : 3		575
VI (i) Spouse & (ii) Siblings^c	1 : 1		582
VII (i) Legit. Parents – per cap. & Siblings – per cap.	1 : 1 : 1 &c	(i) $-\frac{1}{3}$	571
VIII (i) Legitimate Ascendants & Siblings – per cap.	1 : 1 : 1 &c	(i) $-\frac{1}{3}$	571
IX (i) Spouse & (ii) (a) Parents – per cap. & (ii) (b) Siblings – per cap.	1 : 1	(i) $-\frac{1}{2}$; (ii)(a)$-\frac{1}{4}$	582 ; 571
X (i) Spouse & (ii) (a) Ascendants & (ii) (b) Siblings – per cap.	1 : 1	(i) $-\frac{1}{2}$; (ii)(a)$-\frac{1}{4}$	582 ; 571
XI (i) Spouse & (ii) Other kin – 4°	3 : 1		583

^a There is no distinction (as for the legitim) between one descendant and two or more.

^b In this case the spouse has usufruct of $\frac{1}{2}$ the estate if there is one descendant, of $\frac{1}{3}$ if there are more than one.

^c Siblings and half-siblings divide per capita, in the ratio 2 : 1; there is no reserved quota for full siblings.

lawful heirs) in the goods of which the legatee has the usufruct and which are part of the free quota. If they have no bare property rights, then the legacy counts as a simple encroachment and is correspondingly reduced.

There are, finally, certain limitations on a testator's right to dispose of the free quota of his estate. If a person has both legitimate and illegitimate children he may not bequeath to the latter more than would be their share if the inheritance followed the lawful inheritance laws. Unrecognizable children cannot receive

more than half the share given in the will to the least favoured of
the legitimate children. Similarly a second spouse cannot receive
more by will than the least favoured child of the first marriage.
Legal guardians, witnesses to wills and the notaries who draw
them up are all excluded in certain circumstances from receiving
legacies (*Cod. Civ.* arts. 592–600 – where there are further rules
of an even more theoretical interest).

OBLIGATIONS TO GIVE MATERIAL SUPPORT

The laws on the obligation of living persons to give material
support to each other are the counterpart of the laws of succession
(legitim), and I therefore give an account of them here.

The emphasis is on the nuclear family, and, once again, an order
of priority is established which ranks kin, giving, so to speak, the
legal version of a graded intensity of kinship ties. A judge, in
settling how a man shall fulfil more than one obligation shall have
regard to the closeness of the kinship ties, among other things.
The obligations do not lie on collateral kin, but may lie on
affines.

The obligation is to provide *alimenti* – food or nourishment – to
those who are in need and who are unable to provide for themselves (*Cod. Civ.* art. 438). The material support may satisfy a
partial need or supply a total need, and should be in proportion
to the needs of the recipient as well as to the means of the supplier.
'In any case [it] should not exceed what is necessary for the life
of the person in need (*alimentando*) having regard, however, to
his social position' (*Cod. Civ.* art. 438). In some cases a kinsman
can be obliged to supply only what is 'strictly necessary': these
are siblings (*Cod. Civ.* art. 439) and the legitimate or illegitimate
parent of the illegitimate parent of the needy person (*Cod. Civ.*
art. 435). What is strictly necessary, however, may include – in the
case of material support of a sibling aged less than eighteen – the
cost of education. This obligation is not mentioned anywhere else
in this section of the civil code; but the education and instruction
of their own offspring is a duty laid on parents elsewhere.

The material support may be provided either by means of a
food allowance (a special category of wealth which, once in the
beneficiary's possession, is protected against seizure for debt) or
by taking the needy person into the home of the obliged person,
or in any way a judge may determine (*Cod. Civ.* art. 443).

The right to support may be lost if the needy person has been convicted of any of various offences against the morals or honour of a person who would otherwise be obliged (*Cod. Pen.* art. 541): a person who has been convicted for living off his wife's immoral earnings, for example, cannot sue her later for maintenance. The right is lost only against the offended person. The right may also be diminished: a person who conducts himself in a disorderly or reproveable way may have the quantity of his maintenance reduced (*Cod. Civ.* art. 440).

The following is a list of the persons who may be obliged by a court to provide material support. They are identified by their relationship to an *ego* of either sex; the first on the list have the prior duty.

TABLE 22: Alimenti: Order of obligation

Relationship	Cod. Civ. Art.
Spouse	433
Legitimate, legitimized and adopted children, or nearest descendants	433, 436
Adoptive parents	436
Parents or nearest ascendants	433
Illegitimate children	435
Legitimate parent's illegitimate parent	435
Child's spouse	433
Spouse's parent	433
Illegitimate parent's legitimate or illegitimate parent	435
Full siblings	433
Half-siblings	433

It should be noted that absolute priority of obligation may be on a person who is not kin to the needy person at all: that is, on the beneficiary of a gift. This may be anyone, and if he should be a person included in the list above, he moves to the top of the list. The beneficiary, however, is held to give support only to the value of the gift received, or of such part of it as may remain in his possession. The priority is then withdrawn, and the beneficiary reverts to his original position. Affines are released from their obligations when, the spouse having died, the widow or widower remarries; or when the spouse dies and there is no surviving issue of the marriage.

N

People may be released from the obligation at any time if they have not the means to fulfil it; if they have insufficient means they are held to supply as much as they can, and others are called upon. If a person, wholly or partially released on this ground, then enjoys a change in his economic condition, the obligation may be restored. When there are more than one person obliged to give support, each is held to contribute according to his means; this applies when it is a case of several persons in one category (e.g. children); or when the several persons are from different categories. If the several persons are unable to agree on the way in which the obligation should be distributed among them, then a judge may determine.

If one person is obliged to support several needy persons, and is not able to fulfil all his obligations a judge shall make 'opportune provisions' taking into account the varying closeness of the kinship ties between the obliged person and the various needy persons, as well as the latters' possibly different states of need. The judge should also take into account whether or not, if the obliged person were relieved of his obligation, he would be replaced by someone more able than he to take on the obligation.

The law of support, we have already noticed, differs from the laws of succession and inheritance in that it deals with relationships between living persons. It differs also in that the obligations are to supply a determined and relatively fixed amount (what is necessary for the life of the *alimentando*, taking his social position into account); while the inheritance laws are concerned with the proportionate distribution of a variable quantity. It will also be clear that the law of support is, to a far greater extent, a 'crisis' law, one which is applied by a judge in settling disputes, while the laws about legitim and lawful inheritance deal with situations which recur regularly.

The flexibility of the obligations to provide support, and the discretion given to the judiciary may be in part due to the different situations in which the laws (of support and of inheritance) are applied, and of the different nature of what is to be collected together in the one case, and distributed in the other. The laws are not merely similar, however, because they rank kin in accordance with fairly consistent principles. They are also complementary: a person (e.g. an ascendant or spouse) who is excluded from the legal inheritance by the existence of kin who rank 'closer' has, under the law, a right to material support from those who exclude him.

The Cadaster

The Cadaster (*Catasto*) is a land-tax register. It attributes pieces of land (*particelle*) to the person or groups of persons (*ditte*, literally, firms) who are liable to tax on them.

A *particella* is a piece of land which is homogeneous in use (e.g. arable, orchard, or house, etc.), in quality (graded I, II, III, etc.) and in ownership. It must also be in one piece: a field which is crossed by a path is two *particelle*.

A *ditta* is a person or group of persons who are liable to tax. They may be owners, holders in emphyteusis, mortgagors or usufructuaries. The *ditta* is defined by homogeneity of rights: if a man has some mortgaged land, some in emphyteusis, and some by exclusive ownership, he has three *ditte*. Only usufruct can be mixed with other rights in the same *ditta*. When there are more than one person in the *ditta*, the rights must be distributed between them in the same way for each *particella*: if two brothers, Tizio and Caio, have three plots with a different distribution of ownership (e.g. 75 : 25, 25 : 75 and undifferentiated) they have three *ditte*.

The land in the area covered by a Cadaster (usually a commune) is mapped, and the boundaries of all *particelle* are marked, and the *particelle* are numbered. The index to the *particelle* is called the *Stato di Sezione* (or sometimes *particelleria*). This gives references to the Cadaster proper for each *particella*. Another index is the *matricola* which lists all people with taxable rights in alphabetical order, and also gives references to the pages of the Cadaster.

The main Cadaster for Pisticci is a series of some thirty volumes. On each page *particelle* are matched against *ditte*. At the head of the page are names of the individuals who are liable to tax, and the proportions in which rights are shared, if these are specified. Under the names are recorded the *particelle*: the map numbers, the quality, the use of the land, the area of it, and the rateable value. Annotations are made in the margins of how the land was acquired by the *ditta*, and when; and of how it was eventually disposed of, and when; the page number where it came from, and of where it went to, are also recorded. The Cadaster thus provides a record of the movement of land and, by careful use of the index volumes, a tenurial history can be reconstructed for each family and each holding.

The Cadaster is adjusted as each conveyance is registered. This takes some time. Conveyances are registered in the first instance by the notary who makes them; the Registry (*Ufficio del Registro*) then passes the information to the Provincial Registry, which passes the information to the Provincial Cadaster, which in turn passes the information to the local office. In 1965 the Pisticci volumes were up to date for 1962. Changes in ownership are thus recorded currently, but some three to four years later than they occur. Changes in use are not notified to the Cadaster by the people who make them since these are normally improvements and, therefore, carry a greater liability to tax. There are occasional surveys, in theory at five-year intervals, to record changes in the use or quality of land; the surveyors also act as unofficial arbiters in boundary disputes.

The whole Cadaster is renewed from time to time. The current series was begun in 1930; and the Cadaster before that was begun in 1814; its predecessor was the *Onciario* of 1741, which, once compiled, was not brought up to date.[1] By the time the new Cadaster is made the old one is full of inaccuracies, and the new Cadaster makes a clean break with the old one. The numbers of the *particelle* are rationalized, so that the order bears some relation to the distribution on the map; and the rights are recorded as they exist at the time of the survey. No index or concordance bridges the gap between the Cadasters, and it is not, therefore, possible to reconstruct a family history across the gap – or, at least, not unless this is the only task of a researcher with plenty of time. The new Cadasters yield a balance-sheet of the distribution of land within the population at the moment they enter into force; but it is not possible, using manual techniques only, to extract a yearly balance-sheet.

[1] 'La maggior parte dei fondi hanno subito considerabili divarii, . . . Un immensa quantità di fondi sono sorti a rendite opulenti che nel detto catasto non erano annotati . . . I catastuoli creduti atti a supplire a questa mancanza di dati non possono ottenere l'intento . . . Questa gelosa operazione che dispone delle sacre proprietà dei cittadini, e dalla di cui giusta e prudente ripartizione dipende la forza dello stato, e quasi inappellabilmente affidata a pochi briganti cittadini.' G. Zurlo (1802), p. 135.

It was rumoured in 1965 that a new Cadaster was in the offing, and that this would be compiled on punched cards. I hope that the present wealth of detail will be maintained, for then the Cadaster would not only continue to be a source of material for sociologists, historians and economists, but also it would be possible, more or less as a matter of course, to record more carefully and more accurately the important and significant states of affairs and processes which have been reconstructed laboriously and with occasionally irremediable errors in the main text of this book.

APPENDIX V

Equivalent Measures

A hectare is 2.5 acres, approximately.

A kilometre is 5/8ths of a mile, approximately.

A quintal is 100 kilograms, approximately 145 lbs.

A *tomolo* is a local measure with two meanings. As a measure of area it is approximately one acre; as a measure of capacity, approximately 10 imperial gallons, or $1\frac{1}{4}$ bushels.

In 1965 there were 1,740 lire to the pound sterling: one million lire then equalled £575 sterling, In 1973 there were 1,500 lire to the pound: one million lire then equalled £667 sterling.

APPENDIX VI

Synoptic chart of work required to cultivate
one hectare of level land under various crops
(Adapted from M. de Benedictis and M. Bartolelli,
Indirizzi Produttivi, etc., Portici, 1962)

	Grain			Olives			Peas		
	Task	Hours		Task	Hours		Task	Hours	
		M.	F.		M.	F.		M.	F.
Oct.	Fertilize & Sow	13					Plough	12	
Nov.							Prepare furrows, Sow & Fertilize	16	32
Dec.				Harvest (60 trees, 4 qli. oil)	105	610			
Jan.	Harrow & Fertilize	8					Hoc	12	
Feb.				Prune	78		Break soil & Weed		192
Mar.	Apply weed-killer	16					Harvest (45 qli.)		495
Apr.				Fertilize	4				
May									
June	Harvest (15 qli.)	58	5	Plough, Harrow, & Spray	80				
July	Prepare soil (1)	16							
Aug.									
Sep.	Prepare soil (2)	30							

Table Grapes (Mature)			Oranges (Mature)			Tobacco Seed Bed			Tobacco Field		
Task	M.	F.	Task	M.	F.	Task	M.	F.	Task	M.	F.
									Bale up dry leaves	50	50
									Plough	12	
			Harvest (100 qli.)	48	200						
Dry prune	160	80	Fertilize	16							
Fertilize	7					Prepare soil	23				
Spray	120		Prune	200		Sow	4		Plough & Harrow	16	
Plough	20					Weed	43		Fertilize & Harrow	8	
Green prune & Spray	40	150	Plough, Weed & Train branches	368		Fertilize & Spray	21		Transplant & Irrigate	160 / 120	160
Break soil & Weed		150							Weed & Tend plants	160	160
			Irrigate	240							
Harvest (150 qli.)	100	200	Fertilize & Spray	31					Harvest (8 qli.), Thread leaves & Dry	880	800

N*

LIST OF WORKS CITED

ABIGNENTE, G., 1881. *Storia del diritto: il diritto successorio nelle provincie Napoletane dal 500 al 1800* (Nola)

BRONZINI, G. B., 1964. *Vita Tradizionale in Basilicata* (Matera).

COMPAGNA, F. 1963. *La questione meridionale* (Milano).

CROCE, B., 1956. *Uomini e Cose della Vecchia Italia*, 3rd edn, 2 vols. (Bari).

——, 1965. *Storia del regno di Napoli*, 6th edn (Bari).

CUPO, C., 1965. 'Evoluzione passata e situazione presente', in Rossi-Doria, M. and Cupo, C., *Direttrici dello Sviluppo Economico della Lucania* (Bari).

DAVIS, J., 1964. 'Passatella, An Economic Game', *Brit. J. Soc.*, **15**; 191–206.

——, 1969 (a). 'Honour and Politics in Pisticci', *Proc. R. Anth. Inst.*

——, 1969 (b). 'Town and Country', *Anth. Quarterly*, **42**; 171–85.

——, Forthcoming (a). 'An account of changes in rules of inheritance in Pisticci 1814–1960', in *Contributions to Mediterranean Sociology* (J. Peristiany, ed.), (Cambridge)

——, Forthcoming (b). 'Families in Italy', in *Family Structures in Southern Europe* (provisional title) (J. K. Campbell, ed.).

DE BENEDICTIS, M. and BARTOLELLI, M., 1962. *Indirizzi Produttivi e prospettive di mercato per la zona di nuova irri gazione dell'arco ionico. Metaponto.* Part III, vol. 2. (cyclostyle, Portici).

EVANS PRITCHARD, E., 1940. *The Nuer* (Oxford).

FIRTH, R., 1957. *We, the Tikopia*, 2nd edn (London).

GALASSO, G., 1965. *Mezzogiorno Medioevale e Moderno* (Milano).

GRAMSCI, A. 1964. *Quaderni del Carcere, vol. 6: Passato e Presente* (Milano).

INEA (ISTITUTO NAZIONALE DI ECONOMIA AGRARIA), 1947. *La distribuzione della proprieta fondaria in Italia*. 2 vols. and 12 unnumbered vols. of statistical tables (Roma).

ISTAT (ISTITUTO NAZIONALE DI STATISTICA) (ROME), 1960. *Comuni e la loro popolazione 1861–1951.*

——, 1954. *IX° Censimento generale della popolazione italiana* (1951), vol. 1, fasc. 76.

——, 1962. *I° Censimento generale dell'Agricoltura* (1961), vol. II, fasc. 77.

——, 1964. *X° Censimento generale della popolazione italiana* (1961), vol. 3, fasc. 77.

ITALY–PARLIAMENT, 1953. *Atti della commissione parlamentare di inchiesta sulla disoccupazione* (Roma).

KATO, Y., 1965. 'Factors contributing to the recent increase of productivity in Japanese agriculture', *J. Dev. Studies*, **2**; 38–58.

KAYSER, B., 1964. *Studi sui terreni e sull'erosione del suolo in Lucania* (Matera).

LA PALOMBARA, J., 1964. *Interest Groups in Italian Politics* (Princeton).

LEACH, E. R., 1961. *Pul Eliya* (London).

MARX, K., 'The Eighteenth Brumaire of Louis Bonaparte', *Karl Marx and Frederick Engels: Selected Works*, 2 vols. (London), Lawrence and Wishart, 1950.

MINISTERO DELL'AGRICOLTURA DIREZIONE GENERALE DELLA BONIFICA E DELLA COLONIZZAZIONE, n.d. *Quaderni di studio e d'informazione. 8: Notizie, Dati e Documenti sulle strutture fondiarie di pubblico interesse* (Roma).

NITTI, F. S., 1958. *Edizione Nazionale delle Opere di F. S. Nitti,* vols. 1 and 2 (Bari).

PITKIN, D., 1961 'Marital property considerations among peasants: an Italian example', *Anth. Quarterly,* **33**; 33-9

PIZZORNO, A., 1967. 'Familismo amorale e marginalità storica, ovvero perchè non c'è niente da fare a Montegrano', *Quaderni di sociologia,* **xvi**; 247-61.

ROSSI-DORIA, M., 1956. *Riforma Agraria ed Azione Meridionalista,* 2nd edn (Bologna).

——, 1958. *10 Anni di politica Agraria* (Bari).

SALVEMINI, G., 1962. *Il Ministro della mala vita, e altri scritti sull'Italia giolittiana* (Milano).

SALVIOLI, G., 1930. *Storia del diritto italiano,* 9th edn (Torino).

SCARELLA, D. (ed.), 1958. *I Quattro Codici,* 5th edn (Milano).

STIRLING, P., 1965. *Turkish Village* (London).

TEKNE, S. P. A., 1964. *Piano Regolatore Territoriale del Nucleo di Industrializzazione della valle del Basento: Relazione illustrativa del progetto definitivo* (Matera).

VILLANI, P., 1952. 'Il catasto di Carlo Borbone', in *Annali della Facoltà di lettere e filosofia dell'Univ. di Napoli,* II.

——, 1952. 'Note sul catasto onciario e sul sistema tributario nella seconda metà del settecento', in *Rass. Stor. Salernitana,* XIII.

——, 1955. 'Economia e classe sociali nel regno di Napoli (1734–1860) negli studi del ultimo decennio', *Società.*

——, 1962. *Mezzogiorno fra Riforma e Rivoluzione* (Bari).

VOECHTING, F., 1955. *La Questione Meridionale* (Napoli).

WINSPEARE, D., 1883. *Storia degli abusi feodale* (Napoli).

ZURLO, G., 1802. 'Memoria relativa alla Reforma dell'attuale sistema di pubblica economia &c.', ed. by P. Villani, in *Annuario dell'Istituto Storico Italiano per l'età moderna e contemporanea,* **vii** (1955); 133–44.

INDEX

Common Market, 148
Commune, see Government, local
Communion, 10, 61
Communism, Communist Party, 17-
  20 passim, 125; and the Church, 106;
  and Council, 153, 154; and factory
  employment, 150, 153n.; and land
  reform, 146
Communist Alliance of Peasants, 9, 18,
  20, 125, Pl. 1a
Compari, see Godparents
Competition, 15, 55, 155, 156; see also
  Conflict
Confirmation, 61
Conflict, in family, 67, 171-2; at
  harvest, 102-3; between kin, 142,
  159; between neighbours, 67, 70;
  between parent and child, 31-2;
  between parents of intending
  spouses, 30; over resources, 155, 156,
  159; between siblings, 55-7, 124,
  159; between leaders, 153-6
Consiglio Communale see Council,
  town
Contrade, 92, 164-8
Co-operative marketing, 147
Corredo, see Parapherns
Corruption, 17
Council town, 8, 17, 78-80, 84-5, 108;
  see also Government, local; Muni-
  cipio; Political parties; Politicians;
  Politics
Country and town, attitudes to, 2n.,
  9-11, 28, 139-41, 158-62 passim
Courtship, 10, 27-30, 41
Cousin marriages, 25, 66, 123, 143,
  159; see also Kinship; Marriage
Craco, 7
Craftsmen, 12; see also Artisans;
  Labour
Crops, 188-9; see also Food pro-
  duction; Fruit; Grain; Pulses;
  Tobacco
Cupo, C., 11n.

Davis, J., 1on., 22n., 23n., 24n., 56n.,
  65n., 89n., 90n., 94n. 109n.
De Benedictis, M. 99n., 105
Demesne, see Land, public ownership
Demochristian Party, 17, 147, 153, 154,
  155; see also Political parties, Poli-
  ticians. Politics
Dentists, 7

Development programmes, 146-56,
  161; see also Agrarian Reform
  Board
Dialect, vi
Direct Cultivators' Guild, see Catholic
  Peasants' Guild
Distribution of land holdings, see
  Land, distribution
Doctors, 7, 9, 153; see also Health
Dowry, 40, 41, 47, 55, 69, 88; law of,
  172, 173; see also Family; Land;
  Marriage; Property

ENI (Ente Nazionale Idrocarburi), 4,
  152; see also AGIP; ANIC
Education, see Schools; Teachers
Elections, 8; see also Political parties;
  Politics, Suffrage
Employment, see Industrial relations;
  Labour
Engagement, 27-8, 30-1, 41
Ente Communale di Assistenza, 13
Entertainment, 9
Environment, natural, 163-5
European Communities, 148
Evil eye, 35-6
Exchange, 162; see also Gifts; Labour,
  exchange of; Marriage; Property

Factories, 4-5, 13, 42, 43, 149-56
Family, 22-5, 40, 42-72; law, 169-84;
  nuclear, 40, 42-6 passim; see also
  Households; Marriage; Property
Fascism, see Movimento Sociale
Feroleto, 83n., 84, 122, 125
Festivals, 10, 13, 106
Feudalism, 87
Firth, R., 157
Flooding, 165
Food production, 11, 86, 93-103, 117,
  188-9, Pl. 3b
Forestry, 13-4, 84-5, 135
Fragmentation, see Land
Friendship, 64, 65, 158, 162; and
  labour, 11, 100, 101; and politics, 17;
  and visiting, 48, 49; and wedding
  gifts, 36-9 passim
Fruit, cultivated, 96-9 passim, 148;
  wild, 134; see also Crops; Food
  production; Olives; Oranges; Pears;
  Vineyards
Funerals, 10, 55
Furnishings, furniture, 6, 36